M000013642

LIBRARY OF
CONGRESS
SURPLUS
DUPLICATE

How to Profit and Protect Yourself From Artificial Intelligence

Dr. Timothy Smith

Twenty-two
Twenty-eight
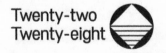

Twenty-two Twenty-eight
Medford, MA
twentytwotwentyeight.com

Copyright © 2018 by Timothy Smith

All Rights Reserved under the Pan-American and
International Copyright Conventions

*This book may not be reproduced in whole or in part, in any
form or by any means, electronic or mechanical, including
photocopying, recording, or by any storage and retrieval
system now know or hereafter invented without written
permission from the publisher.*

Printed in the United States
Published in 2018 by Twenty-two Twenty-eight
First Edition

ISBN: 978-0-9991010-2-5 (paperback)
ISBN: 978-0-9991010-3-2 (ebook)

Library of Congress Control Number: 2018903525

Check us out at:
twentytwotwentyeight.com
Or follow us on Facebook at
Twenty-two Twenty-eight

To Jenn, Rose, and my parents
Dr. James & Elizabeth Smith

Table of Contents

Part 1

How Artificial Intelligence Works

Chapter 1
What Is Artificial Intelligence?

There is an old saying that goes something like this—*If one has a hammer everything becomes a nail.* It seems these days that to artificial intelligence everything is a nail. Artificial Intelligence, or AI, refers to computers that mimic or exceed human capabilities such as decision making, creativity, problem-solving, pattern recognition, purpose, and even consciousness. Nobody makes it through a single day without at least once hearing the words *computer*, *artificial intelligence*, or *robot*. We are surrounded by computers almost everywhere we go these days. People use desktop computers, smartphones, laptops, and tablets at home and at work. Moreover, you will find computers in your car that manage many functions like optimizing engine function, operating anti-lock brakes, and even running the backup camera and bumper sensors. New smart televisions contain a computer to run the screen and manage the multiple inputs of information, including the internet and multiple video streaming sources. To help understand how to profit and protect yourself from artificial intelligence, it is crucial to know what artificial intelligence is and how it works. Before getting to artificial intelligence, it is essential to understand what a computer is, because it forms the basis of artificial intelligence. In the process of understanding what artificial intelligence is and what it is not, we will see that some things are good nails, but not everything is a nail.

What is a computer?

At its most basic level, a computer is a machine that takes in information, performs some operations on that information according to some instructions, and produces new information. The abacus, a simple frame with colored wooden beads on parallel rods, provides an example of a very simple computer. It is a mechanical device invented thousands of years ago that remains in use today. In skilled hands, the abacus can be used to perform various calculations such as addition, subtraction, and multiplication. The beads are stacked up to represent a number. In other words, five beads equal the number five. Using the different columns of beads, large numbers can be added, subtracted, or even multiplied. The slide rule provides another great example of a mechanical calculator. Although replaced now by the electronic calculator, the slide rule since its invention in the 1600s had been an invaluable tool in engineering and mathematics. The slide rule, as the name implies, looks like a stack of different rulers marked in different scales that can be slid back and forth, and a sliding window with a thin vertical line running through it to see the results of the calculation. An accomplished slide rule operator was able to use this device to perform rapid calculations such as multiplication, division, exponentials, and square roots.

Computers today fall into two different classes— analog and digital. Analog computers use continuously changing physical properties such as metal gears, pressure, or electricity to represent numbers and do calculations. The slide rule is a very simple example of an analog computer. The different rulers made of wood or metal sliding back and forth do the calculations. Although not a computer, a good way to think of analog is to think of an old vinyl record. The music information is stored in the cut grooves of the record. When the record is first cut, music vibrates a cutter needle that cuts a groove in a blank record. High frequencies make the needle vibrate back and forth very quickly, and low frequencies move the needle back and forth more slowly. To play the music back on a record, a needle has to move through the groove of a spinning record. Moving through the groove, the needle vibrates just as the cutter vibrated when the record was first made. The music is faithfully reproduced simply by the needle tracking through the bumps in the record grooves. Much more sophisticated analog computers have been developed such as the famous Norden Bomb Sight, which was used extensively by the US Army Air Force during World War II and the Korean Conflict. Using gyroscopes, optics, and mechanical calculations to take into account altitude, wind speed, humidity, and more, the bombsight would calculate when and at which point to drop bombs to hit the sighted target, take control of the aircraft with an autopilot, and drop the bombs at the correct time, "...the mechanical wizard [Norden Bombsight] orders other robots to fly the plane to the right place and drop a bomb at the right time to place it just on the target."[1] The Norden Bombsight represents an engineering marvel and the complexity of tasks analog computing can handle; however, analog computers are built to solve certain questions but lack the flexibility to perform multiple tasks.

The mechanical cash register, which is an analog computer that was once common in every store in the country, has since been replaced by digital electrical cash registers. The digital cash register is only one example of the many types of digital computers in use today. A digital computer is a device that takes in information or data, does some calculations according to a program, and returns the results. Unlike the analog computer that relies on physical properties like gears or marks on a ruler, digital computers use symbols to do their calculations. A digital computer uses binary code to represent numbers and letters. Binary code is a very ingenious way that things like numbers and letters can be represented in a machine. Instead of representing the number '5' as an analog number on a slide rule, the digital representation of the number '5' is a two-symbol or binary system is in the form of a series of eight 0s and 1s. Here are some examples of binary code:

---

[1] Torrey Volta, "How the Norden Bombsight Does Its Job," *Popular Science*, 70, 1945, accessed online September 16, 2017

Character	Binary Code
5	00110101
A	01000001
&	00100110

In binary code, each number from 0-9, each letter of the alphabet, and many symbols such as '&' and '%' have a unique binary code of eight 0s and 1s. The basic binary code needed to represent a number or character is called a *byte*, which is short for the term *binary*. Another kind of binary code is Morse Code. Although only experienced by most people through old movies, Morse Code uses a series of dots and dashes or short and long beeps to represent letters and numbers. For example, in Morse Code 'S' is '...' and the letter 'O' is represented by '---'. The famous mayday distress signal used by ships in trouble is the Morse Code SOS. SOS is sent as '...---...' Digital computers use binary representations of numbers, letters, and symbols to perform logic and calculations to accomplish everything from word processing to modeling weather patterns.

Digital electronic computing is relatively new. The first digital computer was built in 1939. "Beginning in 1935, John Vincent Atanasoff, a physics professor at Iowa State College, pioneered digital electronics for calculating."[2] In collaboration with his graduate student Clifford Berry, Atanasoff built a prototype of the world's first digital computer. They named it ABC for the Atanasoff Berry Computer. Digital computing

---

[2] Berry Atanosoff, "Computer – ABC," https://www.thocp.net/hardware/abc.html Accessed online August 23, 2017

truly emerged just after World War II. The world's first large-scale, general purpose digital computer was built at the University of Pennsylvania and completed on Valentine's Day 1946. Known as ENIAC, or Electronic Numerical Integrator and Computer, this computer commissioned by the US Army to do ballistics calculations ushered in the modern era of large computing systems. To help put it in perspective, ENIAC was massive—weighing 30 tons (more than two city buses) and measuring 60 feet in length. By comparison, the average smartphone in everyone's pocket weighs about 4 ounces and measures 5 inches or so in length, making the smartphone 240,000 times lighter and 144 times shorter than ENIAC.[3] More incredibly, the computer inside the modern cellphone boasts over a thousand times more computing power than ENIAC. After ENIAC, the digital computer came to dominate computing and continues to this day.

How does a digital computer work?

To appreciate the possibilities and limitations of artificial intelligence, it is important to know basically how a digital computer works. At its simplest level, a computer is an electronic machine that takes in information, places that information in memory, performs calculations, and then outputs new information. The input information can come from typing on a keyboard, an electronic sensor like a microphone or a digital camera, or the output from another computer. A

---

[3] "The History of Computing, ENIAC," https://www.thocp.net/hardware/eniac.htm, accessed online on August 19, 2017

What Is Artificial Intelligence?

computer has four main components:[4]

◊ Central processing unit or CPU—the CPU is the part of the computer that performs the calculations and logic—sort of the brains of the operation. It will perform these tasks with lightning speed. The faster the CPU, for the most part, means a faster computer.

◊ Primary Memory—the place for the CPU to very quickly access the data and instructions it needs to do its job. The information in primary memory is not permanent and disappears when the computer is turned off.

◊ Secondary Memory—the place where information such as photos and documents are permanently stored.

◊ Input and output devices—Input devices convert information into the digital form. Some typical input devices: keyboards, digital cameras, and microphones. Output Devices convert information back into a form people can work with. Some typical output devices: speakers, printers, and monitors.

The basic process of computing centers around data inputs, calculations, and data outputs. Taking a picture with your smartphone or digital camera provides a practical example of the steps in how a computer works. Beginning with input, the camera captures the image on a light sensing chip, and the phone's computer (CPU) transforms electrical signals from the

---

[4] Charles Severance, Python for Informatics, Exploring Information, version 2.7.3, 2009, Accessed on line August 21, 2017

chip into numbers. The numbers represent a kind of numerical map that describes the colors and shadows in the picture.[5] Think of color-by-numbers books. Each number on the picture represents what color pen to use to fill in the image. The same holds for a digital image, except there are thousands of tiny dots and thousands of colors for the computer to use. The picture, now a map of numbers, is stored in secondary memory. To view the picture on the screen of the camera, the picture file must be read from the secondary memory by the CPU to convert the digital map into the electrical signals that make the screen display the photo. Another example would be editing a digital picture on a smartphone. Perhaps the people in a picture look terrible because they all have red-eye from the flash. The red-eye tool easily fixes the problem. With the digital picture as input to the secondary memory to the smartphone's computer, the red-eye software tells the computer to do some calculations. The smartphone CPU stores the picture in primary memory and does some mathematical changes to the picture, replacing the red with black and finally outputs the modified picture to your screen with the red-eye removed.

The memory, CPU, and input/output devices, in other words, the *hardware*, only make up part of a computer. The computer also needs instructions on what to do. A computer merely follows orders and will do so very diligently, but it must be told clearly what to do. A computer receives its orders from a program which is also called an *algorithm.*

Now more than ever, the word 'algorithm' flies

---

[5] "Digital Cameras," http://www.explainthatstuff.com/digitalcameras.html, accessed online Sept 2017

freely about in the news, advertising, and conversation. Google Ngram tracks the usage of words in books over time, starting with books published hundreds of years ago through the present day. Ngram demonstrates the increase in popularity of the word 'algorithm' in books starting in the early 1960s and continuing to increase through today. Even more explosive growth in the usage of 'algorithm' appears in Google searches when combined with other terms such as Tinder, Instagram, or Bitcoin. Google Trends, which tracks what people search on the internet, shows huge increases in searches containing the word 'algorithm,' but what does the word really mean? In short, *algorithm* stands for a set of step-by-step instructions designed to do something.

The word algorithm comes from the name of the Persian Muslim scholar, philosopher, and mathematician named Muhammad ibn Mūsā al-Khwārizmī who lived from ~780-850 CE.[6] He wrote great books on mathematics, and he introduced the decimal to the Western World through a translation of his works. The translation latinized his name to *Algoritmi*. We often hear the term *algorithm* associated with computers, but algorithms do not only exist in computers. People use step-by-step instructions all the time to solve a problem or get something done. For example, to roll the change accumulated in a change jar, a person will first separate all the coins by type—pennies with pennies, nickels with nickels, etc. After sorting the change by type, the coins must be stacked into the amounts for each roll such as 50 pennies or 40 quarters per roll. Once the

---

[6] "Abu Ja'far Muhammad ibn Musa Al-Khwarizmi," *MacTutor History of Mathematics archive,* accessed online on January, 19, 2018, www-history. mcs.st-andrews.ac.uk

stacks of coins have the correct number, the coins must be placed in the proper paper rolls such as pennies in the penny rolls. In short, rolling coins uses an *algorithm*, or specific steps designed to complete a particular task. Following the instructions to make bread constitutes a more complex non-computer algorithm. The baking algorithm requires the use of precisely measured, specific ingredients such as flour, yeast, salt, and water. Moreover, the bread baking algorithm requires the baker to mix and knead the ingredients, wait for the bread to leaven, and then bake the bread at a specific temperature and for a certain amount of time.

Computer algorithms also perform particular step-by-step tasks to achieve a specific goal. Sometimes the algorithm may be straightforward, such as finding the biggest number in a long list of numbers or counting the amount of time a particular word like *cat* gets used in a document. In the first algorithm, looking for the largest number in a long list of numbers, the computer will be instructed to look at the first number in the list, call it the largest number, and compare it to the next number in the list. If the next number that comes up is larger, then the program will call that number the largest number. The algorithm will fly through the entire list very quickly and read out the largest number. For the word counting example, the algorithm starts with a zero for the number of times it finds *cat*. It will search from front to back through the document looking for the word *cat*, and every time it finds *cat*, it will add one to the number. Such algorithms sound simple, but they can save a lot of time and be very accurate. Other algorithms perform more complex tasks such as the search algorithm that made Google the world's largest internet search engine. Google developed and patented

What Is Artificial Intelligence?

PageRank. PageRank helps Google provide the best search results when searching through the trillions of web pages on the World Wide Web. Basically, to do this, PageRank looks at the search words you type into Google and finds all the pages containing your search. Next, PageRank searches through the pages it found and chooses the ones that have the most links connected to them. PageRank uses the idea that good quality internet pages will have more links to them than other pages with low-quality information. The highest scoring pages will be at the top of the Google search results. Using the PageRank algorithm, Google solved the problem of finding the best search results across trillions of pages in a stepwise manner.

The great Persian mathematician, Muhammad ibn Mūsā al-Khwārizmī, provided the basis for the word algorithm. Although *algorithm* sounds hard to interpret, it stands for a step-by-step process designed to solve a problem. People use algorithms in their everyday lives without touching a computer. Baking bread and sorting laundry or coins employ algorithms. Computers use algorithms to solve particular problems such as counting words or returning the best search results. Computers surround us today, and therefore algorithms large and small bring us our email, route our phone calls, make our Facebook pages work, and, in other words, support many of the things we do every day.

An algorithm is how people talk to digital computers and tell them what to do. An algorithm tells the computer what to do with the information coming in, how to process or crunch that information, and then what to do with what comes out. For example, the program can tell the computer to save the output information or to send that information to a printer. Computers need data and need algorithms to know what to do with that data. Computer scientists write computer algorithms. Writing computer algorithms often is just referred to as *coding*, because all algorithms are written in code. Code comes in specific computer languages with names such as Python, Java, and C++. Each language (and new ones are being developed all the time) is designed to work with different types of computers and to perform various kinds of tasks.[7] Some languages are very general, and others may be optimized to run on a particular kind of machine. According to Spectrum IEEE (Institute of Electrical and Electronics Engineers), Python rose in 2017 to become the top programming language.[8] Python, known as a general-purpose programming language, is popular for coding data analysis tools, animation software, and artificial intelligence. Another prevalent programming language called Java helps programmers build web applications. C++ (pronounced "see plus plus") is used to program many software applications on desktop computers, servers, and even telephone switches.

Computer languages, just like spoken languages, have particular words and rules that the programmer must follow so that the computer can understand the commands and perform its tasks. The rules are very strict, and if the programmer does not follow the rules or if the incoming data is not in what the computer program is expecting, the computer will stop and throw

[7] Marc Andreessen, "Why Software Is Eating the World," *Wall Street Journal*, August 20, 2011
[8] "The 2017 Top Programming Languages," Spectrum IEEE, *https://spectrum.ieee.org/computing/software/the-2017-top-programming-languages*, accessed on September 23, 2017

an error. The same thing would happen with two people speaking English; if a friend calls and asks, "What time would you like me to come over for dinner?" And you replied, "Falcon?" She would say, "What? You are not making any sense!" You were expecting a time but heard the name of an animal instead. You threw an error. Computer languages use exact terms and rules, and when the computer code does not make sense because of bad logic, grammar, or punctuation, the computer calls out an error. Many computer programs are written to execute a specific task and can do so extremely quickly—much quicker than a person—but the computer will perform the duties with no real understanding of what it is doing. As in the analogy from human interaction before, you expect your friend to give you a time to meet, and her strange response will probably make you ask her if she is all right. Hearing your friend speaking strangely may cause concern because that could be a sign of some other problem. In contrast, computers will be literal and just throw an error. Computer programs can become extraordinarily complicated and be designed to anticipate mistakes and respond to them, but they are still only performing the tasks as assigned.

The Difference Between Artificial Intelligence and Traditional Computer Programming

The working definition of *artificial intelligence*, or *AI*, in this book refers to computers that mimic or exceed human capabilities such as decision making, creativity, problem-solving, pattern recognition, purpose, and even consciousness. To make sense of the recent explosion of artificial intelligence, it is essential to understand that artificial intelligence grows from a new type of computer programming. In the first chapter of "Difference Between Artificial Intelligence and Traditional Methods," Henk van Zuylen separates traditional computer programming from artificial intelligence by the type of problem the computer program solves.[9] Traditional computer programs contain all the logic and mathematics needed to solve a particular problem. In other words, the programmers must anticipate every aspect of a problem and provide explicit instructions for what the computer must do. Whereas artificial intelligence will learn and adapt in ways the computer scientists did not code into the system. Constance Zhang put it very well when she wrote, "Learning is the process of converting experience into expertise or knowledge. A learning algorithm does not memorize and follow predefined rules but teaches itself to perform tasks such as making classification and predictions through the automatic detection of meaningful patterns in data."[10]

Making a peanut butter and jelly sandwich provides a very entertaining example of traditional or *explicit* programming that computer science teachers will often use to help first-time students understand how precise explicit programming needs to be.[11] The example requires the class of students to develop a set of rules to guide the teacher who will act like a robot. The teacher acting like a robot will follow the rules

[9] Henk van Zuylen, "Between Artificial Intelligence and Traditional Methods, Artificial Intelligence Applications to Critical Transportation Issues," *Difference*, 2012 Accessed online on September 21, 2017

[10] Constance Zhang, "Infusing Machines with Intelligence - Part 1," *Platinum, 4 November 2016*

[11] "Making Peanut Butter and Jelly," Duke University, https://www.cs.duke.edu/courses/spring14/compsci101/assign/01_algorithms/pbj.php, Accessed on line August 14, 2017

explicitly to make a peanut butter and jelly sandwich. The exercise can be very instructive and funny at the same time when the teacher uses real bread, utensils, peanut butter, and jelly. The peanut butter and jelly sandwich example provides a great way to understand the difference between traditional computer programs and the computer programs behind artificial intelligence. The teacher challenges the class to think of all the steps it takes to go from the concept of a peanut butter and jelly sandwich to producing an edible one and then *code* some rules to guide the teacher.

To make the point very clear, let us walk through the primary steps involved in making a peanut butter and jelly sandwich. First, the concept of a peanut butter and jelly sandwich needs to be understood. For our discussion, we will define a peanut butter and jelly sandwich as two pieces of bread with peanut butter spread evenly on the face of one slice of bread and jelly spread evenly on another slice of bread, and the two slices of bread are pressed together so that the peanut butter comes in contact with the jelly. At first glance, the operation sounds like a reasonably simple task until you just dig a little bit deeper. To program a computer to run a robot to make this simple sandwich, you must also consider all the elements a person takes for granted as merely common sense. What is peanut butter? How does the robot know what peanut butter is? Can it see by the label? How does it open the jar? Can it confirm that the jar labeled peanut butter contains peanut butter by smell, sight, texture? The same goes for jelly, bread, and the utensils needed to do the task. The instructions must be extremely explicit about all the actions required to successfully make a robot to do all of this very precisely.

With all this in mind, the teacher will follow the sandwich-making instructions from the students *explicitly*. The instructions acting as a program or code from the students may sound correct but may be too vague or imprecise. Take for example the direction "put the peanut butter on the bread." The teacher may just pick up the jar of peanut butter and set it on top of the bread. An engaged class will laugh but also will need to add instructions to open the jar of peanut butter and scoop out the peanut butter. The instructions may still say "Put the peanut butter on the bread but spread it all over it instead of just one side." Another instruction may be, "Put the two pieces of bread together." The teacher can push the two sides together, and the class will see that you need to use precise instructions to put the side with the peanut butter against the side with the jelly to finish the sandwich. This simple exercise demonstrates how explicit programming works. The computer does not need to learn what peanut butter is or what a sandwich is as long as all as long as the program contains all the necessary instructions. In other words, the computer only works when preloaded with all the instructions it needs.

In contrast to traditional programming, artificial intelligence programming makes the machine do the learning without explicit instructions. Artificial intelligence acts more like a human, because it learns like a human learns—by example and repetition. Instead of the programmer designing every rule the computer needs to function just like in the peanut butter and jelly example above, the programmer creates algorithms that are capable of learning. The problem of using machines to sort vegetable and fruit produce nicely illustrates the difference between explicit programming and artificial intelligence. A sorting device could be built to sort tomatoes at a factory using either explicit programming or artificial intelligence. Imagine the device can sort tomatoes into three categories—the unblemished ones for selling fresh at the market where they will fetch the highest price, the damaged ones to be separated for dicing and canning, and the rotten tomatoes for the compost heap. A very skilled computer programmer working with a robotics engineer would design the sorting machine with precise instructions to sort tomatoes, including a color test and maybe a test for roundness to sort the market ready tomatoes from those destined for canning or even the trash. The explicit program would have all of its instructions up front as to how to recognize and evaluate tomatoes and only tomatoes. However, when tomato season ends, the explicitly programmed sorter would just have to sit idle until the tomato harvest next year. Now a programmer and an engineer could go back to work writing a new explicit program to reprogram the machine to help with the apple harvest later in the fall, and they would need to make rules that define what a good apple for sale is versus those damaged ones better suited for apple juice or the compost. The explicit programming will be very time consuming and still not prepare the sorting machine for the Brussels sprouts after the apple harvest or the cherry harvest next summer.

Using artificial intelligence, the computer programmer codes the computer controlling the sorter to learn from pictures of good, damaged, and rotten tomatoes. Instead of writing explicit computer code that defines tomato quality, with artificial intelligence the computer learns on its own. The machine *looks* at thousands of pictures of tomatoes that people have labeled as suitable for market, canning, or the compost. The computer incorporating artificial intelligence makes up its own rules of how to classify tomatoes instead of explicit programming defining all the rules up front. The tomato sorter will be capable of sorting tomatoes based on the rules it made up through observation. Furthermore, because the computer can learn through artificial intelligence, it can just as easily be taught to determine the quality of other types of produce. An artificially intelligent sorting machine would be much more versatile than one requiring traditional programming. With all the different crops that mature throughout the year, there would be almost no downtime for the AI sorter, which makes the machine much more profitable too.

Computer vision offers another fascinating example of how artificial intelligence has changed what computers can now achieve. Artificial intelligence has dramatically accelerated the development of computer facial recognition and more generally of giving computers the fantastic ability to see. In the *Handbook of Facial Recognition*, Stan Z. Li and Anil K. Jain describe a variety of approaches computer scientists have used over the years to develop computer facial recognition. Earlier methods used explicit programming to define recognizable features of faces using specific points such as nose, eyes, and chin to make a map of the face that can be measured. Features such as the distance between the eyes and length of the nose can then be used to search a database of photographs with known faces for a match. This technique works well when the faces and pictures line up well, but some photographs have a head tilt, lighting differences, and partial faces making explicit mapping programs much less effective than a person at recognizing a face. However, with artificial intelligence, computer facial recognition has dramatically improved. In fact, the Chinese company Baidu published a paper that claims that their artificial intelligence enabled facial recognition system called Facial Recognition via Deep Embedding has a 99.7% accuracy in facial recognition, which is better than a human.[12] Artificial intelligence gives computers for the first time the ability to very accurately see and recognize faces even more accurately than people do.

Artificial Intelligence, or *AI*, refers to computers that mimic or exceed human capabilities such as decision making, creativity, problem-solving, pattern recognition, purpose, and even consciousness. Artificial intelligence has only become possible with the rise of fast, powerful digital computers. Computer scientists have developed programming languages and techniques that give machines the ability to learn in a way similar to how people learn. In other words, computers do not need to be explicitly programmed; instead, through observation, computers now develop their own rules and methods of solving problems such as sorting, facial recognition, and many more problems previously only thought solvable by people.

---

[12] Jingtuo Liu, Yafeng Deng, Tao Bai, Zhengping Wei, Chang Huang, "Targeting Ultimate Accuracy: Face Recognition via Deep Embedding," citation arXiv:1506.07310, 2015, accessed online on September 26, 2017

Chapter 2
The Rise of Artificial Intelligence

The terms *artificial intelligence* (AI) and *robot* conjure up images from science fiction books, comics, and movies of computer-run machines that are vastly more intelligent and powerful than flesh and blood humans and the rest of the vulnerable animals and plants of our world. The robot army in the *Matrix* movies ruling over a grim world cast in hellish darkness with humans being used as comatose slaves to produce energy typifies a frightened skeptical view of the future. *The Matrix* movies depict a future when computers become intelligent and humans increasingly obsolete. To understand and prepare for a future with more intelligent machines, the terms and concepts of *AI* and *robot* need to be clarified. Chapter 2 will work to explain these terms and give you an idea of how artificial intelligence and robotics work. Knowing what they do and how they function will help you to understand the exponentially growing area of artificial intelligence, information technology, and robotics. AI, information technology, and robotics are changing our world in different ways from manufacturing and commerce to entertainment, education, healthcare, warfare, human interaction and more. The concept of artificial intelligence is the concept of machines being able to perform tasks historically only associated with human capabilities such as reasoning, reading, writing, listening, speaking, seeing, planning, making goals, and then taking action to achieve those goals. Chapter 1 provided examples of artificial intelligence empowering computer vision through facial recognition and decision making through sorting.

A Brief History of Artificial Intelligence

It appears that artificial intelligence has burst onto the scene in recent years with computers defeating humans at games like chess and *Jeopardy!* as well as many big companies such as General Electric and Google advertising their transformation into artificial intelligence companies.[1] However, the notion of artificial intelligence in many respects is not new. In the excellent book, *Machines Who Think*, Pamela McCorduck details how history is full of examples of man seeking to imitate his thinking and action in inanimate objects. People have played with the notion of inanimate objects coming to life for thousands of years. References to clay men go back deep into early folklore. For example, Jewish folklore mentions a human-like being often hostile towards all but its creator called a golem. A

---

[1] Sean Captain, "GE Wants to Be the Next Artificial Intelligence Powerhouse," *Fast Company*, 2016, accessed online September 28, 2017

notable golem story comes from the 16th century where the Rabbi of Prague, Judah Loew ben Bezalel, creates a mighty golem to defend the Prague ghetto against anti-Semitic attacks. He named the golem Joseph Golem and also employed the golem keep the temple clean. However, the golem eventually got out of control and had to be destroyed.[2] Making the inanimate alive and intelligent is a recurrent theme of children's literature. Walt Disney adapted the 1881 Italian children's book by Carlo Collodi that tells the tale of Pinocchio. In the movie, Pinocchio, a mischievous wooden marionette, is animated by a fairy to keep the woodcarver, Geppetto, company. In the end, through trials of character and again through magic, the wooden boy is rewarded with becoming not just animated but fully human. More darkly, Mary Shelly's Frankenstein from her novel *Frankenstein, or The Modern Prometheus*, published in 1818, depicts a monster constructed and brought to life by Dr. Frankenstein.[3] The monster, like the golem, cannot be controlled and eventually must be destroyed. Since *Frankenstein*, the result of humans tampering with life or creating artificial intelligence and robots as depicted in art and literature often ends poorly, cautioning man not to play with fire he cannot control. From the rogue computer, Hal, in *2001: A Space Odyssey* to the tyranny of Skynet in the *Terminator* movie series, humans often become redundant and powerless against their creations. However, with all the warnings and dark scenarios, researchers continue now more than ever to develop artificial intelligence.

---

[2] Pamela McCorduck, *Machines Who Think*, A K Peters, Ltd. Natick, Massachusetts, 2004, accessed online August 2017
[3] Mary Shelly, *Frankenstein; or The Modern Prometheus*, 1818

Earlier in the 20th century, building on the fields of mathematics and formal logic, Alan Turing theorized that machines could perform any mathematical or logical function by merely using symbols much in the way a digital computer works. Turing was born on June 23, 1912, in London and studied mathematics at Cambridge University and later earned his Ph.D. at Princeton University. At the age of twenty-four, Turing published a ground-breaking paper titled, "On Computable Numbers, with an Application to the Entscheidungsproblem [decision problem]"[4] in which he remarkably breaks down into simple steps how humans perform calculations.[5] In the same paper, Turing proposes how machines could perform these calculations, which formed the basis for digital computing. The movie *The Imitation Game* featured Turing and dramatized his role in cracking the German's secret code, Enigma, which profoundly turned the tide of World War II. Turing also theorized about intelligent machines. In the early 1950s, he suggested that for a computer to be truly intelligent, it would have to be able to hold a conversation with a person, and the person would not be able to tell she was having a conversation with a machine. The test is now referred to as the *Turing Test* and is still a benchmark for intelligent machines today.

The term *artificial intelligence* first appeared in the mid-1950s and was introduced by John McCarthy

---

[4] Alan Turing, "On computable numbers, with an application to the Entscheidungsproblem," *Proceedings of the London Mathematical Society*, Ser. 2, Vol. 42, 1937, accessed online September 17, 2017, http://www.turingarchive.org
[5] Chris Bernhardt, *Turing's Vision: The Birth of Computer Science*, 2016, MIT Press

while he was an assistant professor of mathematics at Dartmouth University in Hanover, New Hampshire. McCarthy, along with a small handful of men, started the field of artificial intelligence at the famous Dartmouth Conference in the summer of 1956. These men included Marvin Minsky, then a Harvard Junior Fellow in mathematics and neurology and later the co-founder of the Artificial Intelligence Lab at MIT, Herbert A. Simon, who would later win the Nobel Prize in Economics, and Allen Newell, a researcher in computer science and cognitive psychology at the RAND Corporation. McCarthy believed that gathering like minds in one place without distraction would lead to great advances in the machine intelligence.[6] The Dartmouth Conference set in motion the first great wave of research in AI. McCarthy and the other conference goers believed that machines could be programmed to perform tasks that previously only humans could do, which is the essence of AI. The Dartmouth Conference brought together some remarkable people who continued to develop artificial intelligence throughout their careers. McCarthy would go on to create a computer language called Lisp for programming artificial intelligence and robotics.[7] The field of artificial intelligence grew throughout the 1950s and 60s but would slow down in the 1970s as financial backers pulled out due to a lack of success. That time is known as the *AI Winter*. Remarkably, many of the theoretical foundations for the explosion of artificial

---

[6] Pamela McCorduck, *Machines Who Think*, A K Peters, Ltd. Natick, Massachusetts, 2004, accessed online August 2017
[7] Martin Childs, "John McCarthy: Computer scientist known as the father of AI," *Independent*, November 2011, accessed on line October 1, 2017

intelligence we see today originated in the 60s and 70s, but the computer hardware was too slow, and the supply of data was too small to develop artificial intelligence.

With the advent of faster, more powerful computers, better and cheaper memory, and vast amounts of data, the past ten years have seen the remarkable rise of artificial intelligence. A significant part of the growth of artificial intelligence has been the availability of big data. For decades, certain types of artificial intelligence were not possible because the computers were too slow, did not have enough memory, and there just was not enough data to train the machines. All of that changed in the past ten years with the advent of faster computers, cheaper and more abundant memory, and the arrival of big data. The term *big data* gets used in the news and commercials all the time.[8] At its core, big data refers to massive data sets that cannot be analyzed by traditional methods. Traditional methods may refer to using spreadsheets or traditional databases that handle thousands of data points or even millions of data points. A conventional data analysis may be looking at polling for a presidential election and may contain a million (1,000,000) data points. Big data can refer to data in the terabyte, petabyte (1,000,000,000,000,000) range or even more.

It can be hard to visualize what a terabyte is. Thinking of a terabyte like sand on a beach may provide a more definite sense of the amount of data that big data represents. The following is a very basic estimation of big data using sand as an analogy for data points.

---

[8] Victor Mayer-Schoenberger and Kenneth Cukier, *Big Data: A Revolution That Will Transform How We Live, Work, and Think*, John Murray, 2013

The Rise of Artificial Intelligence

Remember from Chapter One that a byte of data is the chunk of information that represents the smallest piece of information that a computer needs to describe a letter like *A* or a digit like *1* or *2*. The amount of data that a computer consumes in solving a problem or in playing a song on iTunes is often described in some multiple of bytes that we often hear like kilobytes which about 1000 bytes or megabytes which is 1000 times more. For example, a typical song on iTunes uses about 4 megabytes of memory, or 4 million bytes! That sounds like a lot of bytes. So, let's put it in perspective of sand. Take a child's beach bucket; if you fill it with sand, it will contain about 9 million grains of sand or just over two songs on iTunes. If the sand were bytes of data, it would be about 9 megabytes of data in the bucket. Now imagine standing on a beautiful sand beach as long as a football field and 50 yards to the water. That beach contains roughly 926,000,000,000,000 grains of sand or just under one terabyte of sand. Big data works with data in the terabyte range and even the larger petabyte range, which is vastly more massive than an iTunes song or found in a typical spreadsheet. The difference between ordinary data and big data is billions and billions and even billions and billions and billions and billions.

The term *artificial intelligence* was coined in 1956 at The Dartmouth Conference. Following that conference, computer scientists have developed artificial intelligence algorithms that could learn on their own, but only until recently with the availability of faster, more powerful computing and enough data has artificial intelligence arisen.

Narrow Artificial Intelligence v. General Artificial Intelligence

The term *Artificial Intelligence* refers to a computer that can learn, adapt and develop knowledge and skills on its own beyond the original instructions given by its creator. The tasks are historically only associated with human capabilities such as reasoning, reading, writing, listening, and speaking, planning, having goals, and then taking action to achieve those goals. The explosion of artificial intelligence applications has already produced some remarkable results and has brought a distinction in the field of artificial intelligence between different types of artificial intelligence—*narrow* artificial intelligence and *general* artificial intelligence. Narrow artificial intelligence, sometimes called weak artificial intelligence, describes the AI that does one kind of task very well such as facial recognition, translation, or sorting. However, narrow AI cannot change its mind and work on some other problem. A person effortlessly shifts from sorting some objects like ripe tomatoes from green ones to having a conversation with a friend who is having a problem with his neighbor. People apply their intelligence freely to problems at hand as they arise. This type of intelligence that people have is known as general intelligence. General intelligence in the world of computing is called *artificial general intelligence*, or AGI. AGI is is also referred to as strong artificial intelligence. AGI goes beyond narrow artificial intelligence. AGI would be the case of an entirely independent, *conscious* entity, fully capable of surviving and thriving on its own, an entity with its own will, motivations, and would live among us. All of the applications of artificial intelligence deployed today fall into the category of narrow artificial intelligence.

Narrow Artificial Intelligence

Narrow artificial intelligence covers a wide variety of different specialized activities from voice recognition to fraud detection. Even driverless cars use a combination of narrow artificial intelligence to navigate vehicles safely along roads and around parking lots. Now remember from Chapter One, artificial intelligence programming allows the computer to learn itself without explicit instructions. The kind of artificial intelligence that learns by itself is often referred to as *machine learning*. Machine learning artificial intelligence acts like a human because it learns as a human learns—by example and repetition. Types of narrow artificial intelligence known as machine learning form the basis of many artificial intelligence applications used today in both the public and private sectors. The following section of the book will detail four types of machine learning:

◊ Supervised

◊ Unsupervised

◊ Reinforcement

◊ Deep

The four types of narrow artificial intelligence provide a general sense of how narrow artificial intelligence works and the areas of application at which it is the best.

Supervised Machine Learning

Have you ever wanted to predict the future? Artificial intelligence can help with that using a technique known as *supervised machine learning*. Supervised machine learning, as the name implies, refers to the area of narrow artificial intelligence in which machines *learn* from known examples. With supervised machine learning, the computer can practice predicting the future with information that it already knows is correct. In a way, it is the same as working on problems that you have the answers for already. The set of known answers is called a *training set*. The computer, beginning with the training set, looks for ways to predict the right answers. This training is called supervised machine learning. For supervised machine learning, the computer needs people to define what it is learning, in other words, to supervise its learning. Supervised learning uses information from the training set to classify or sort information. The sorting and classifying reveals patterns in the information that the computer can use to predict the future. In other words, supervised learning is very good at answering the question, "Is it A, or is it B?" By extension, it is also great at the question, "Will it be A or B?" or "Will it be A, B, or C?" "What will customers want in the fall? Will they want vanilla, chocolate, or coffee ice cream?" Supervised machine learning can sort, and it can predict.

Here are some great questions for supervised machine learning:[9]

---

[9] A. Criminisi, J. Shotton, and E. Konukoglu, "Decision Forests for Classification, Regression, Density Estimation, Manifold Learning and

◊ What will bring in more return customers, a 20% discount coupon or a $20 instore cash card?

◊ What will sell more this season, the light denim jeans or the dark denim jeans?

◊ What are the objects in these photographs?

◊ Will my marketing campaign be more successful in print or on the radio?

◊ How much will that house cost depending on its distance from a commuter rail stop?

◊ What was the mood of the tweets around my commercial during the Super Bowl?

◊ Will this client go to my competitor?

The last question in the list above can also be rephrased to, "Can I predict the customer churn for my company this year?" *Churn*, or churn rate, in business means the number of subscribers or clients that discontinue their service contracts over time. Successful businesses grow by signing up new clients while keeping their churn rate low. Researchers at Microsoft in their machine learning division called Azure authored a *white paper*, or authoritative business report, titled, "Analyzing Customer Churn by using Azure Machine Learning."[10] In the white paper, the researchers demonstrated that they could better predict customer churn in the highly competitive mobile phone

---

Semi-Supervised Learning," Microsoft Research technical report TR-2011-114, p 15,
[10] https://docs.microsoft.com/en-us/azure/machine-learning/machine-learning-azure-ml-customer-churn-scenario, accessed November 27, 2016

industry using machine learning than using traditional measures such as comparing phone cost, data plans, and reception quality between carriers. Such traditional factors do not accurately predict which customers will leave for another carrier. The authors developed and tested a model that used machine learning and data collected directly from customers to get a better prediction of customer churn. The data used in the model included how and when customers use their phones, duration of calls, data consumption, number of data overages, and customer support data. Supervised machine learning turned this data into predictions about which customers were most likely to jump to another carrier. The Azure team found that their model outperformed more conventional methods in identifying the highest flight risk customers. Knowing which customers pose the highest risk of moving to another carrier gives management a chance to direct special attention to those customers. Such concierge service based on supervised machine learning may address the needs of flighty customers and successfully retain business.

Figure 1 is an example of a decision tree that provides an answer to the question "Should I go play outside?" It uses aspects of the weather such as humidity, outlook, and amount of wind to answer the question. By following the tree and answering questions at each of the branches, one comes clearly to the decision whether or not to go play outside. Decision trees form the basis of a number of supervised machine learning tools.

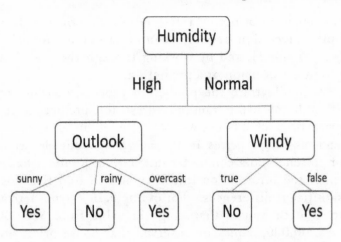

Figure 1. Decision Tree[11]

Different supervised machine learning algorithms have names like logistic regression, bagged trees, and random forest. *Random forest* is the name of a supervised machine learning tool that builds decision trees from data to classify information and make predictions. Just like the graphical tree above, random forest takes data and learns what information delivers the best answer. Since random forest is computerized, it can build thousands of decision trees at high speed and compare the different trees to find the most informative splits in the information. The algorithm will build the trees randomly. The random building of the trees lets the algorithm try many solutions that perhaps an expert may not think is relevant. That randomness helps remove bias from the prediction. Random forest will try thousands of scenarios and use statistics to find

---

[11] Photo Source: Wikipedia Commons, https://en.wikipedia.org/wiki/File:Decision_tree_for_playing_outside.png

the most informative trees out of the thousands it builds. Decision trees divide information into branches (see Figure 1), and by working through the branches, produce a decision or a prediction.

To illustrate the prediction process, consider an example of using random forest for predicting the price at which a house would sell based on information such as house prices in the neighborhood, elevation, or closeness to a commuter rail station. Random forest splits the information into branches to find the most significant differences. For example, in a hypothetical town with houses that range in value from $70,000 to $200,000, it may be observed that house prices are always over $110,000 when they are a mile or less from a commuter rail station. The tree can use other information too such as the number of bathrooms in a house, availability of off-street parking, the distance from a school, or whether the house has municipal sewer and water or a well and septic system. Random forest will build many decision trees and then figure out which trees produce results closest to the training data. Since random forest is computerized, it can build thousands of decision trees and compare them to find the most informative patterns. Once random forest learns the best pattern, it can use that decision tree to predict house prices of other homes in the neighborhood.

Another example of the application of random forest can be found in Microsoft's Xbox 360. Xbox 360, the powerful and popular video gaming system, achieved great success in computer vision using random forest to power its Kinect system.[12] The Kinect system

[12]J amie Shotton, et al, "Real-Time Human Pose Recognition in Parts from Single Depth Images," http://research.microsoft.com/pubs/145347/

captures game player body movement, which allows players to physically interact with their video games. For example, "Kinect Adventures" captures players' full body motions to play games such as river rafting and track and field. Many popular dance games use Kinect such as *Just Dance*. *Just Dance* combines popular songs with challenging choreography to immerse the game player in a realistic dance competition experience. Kinect comes with a built-in random forest that has been trained to identify thirty-one body parts such as left hand, left elbow, and left shoulder. Using the decision tree built into Xbox and a 3D camera, Kinect builds a 3-dimensional map of a gamer's body, and using random forest, the computer determines continuously where all thirty-one of the body parts are located on the map. After the map is constructed, the computer in the game console continuously uses decision trees to track body movements in real time to provide the physical interaction with the games.

Cardiovascular disease causes heart attacks and strokes. According to the World Health Organization in its 2017 fact sheet, cardiovascular disease kills 17.7 million people per year worldwide.[13] That number accounts for 31% of all deaths, which makes cardiovascular disease the number one killer. In light of the lethality of cardiovascular disease, doctors pay very close attention their patients for signs of potential heart attack and stroke. The American College of Cardiology/American Heart Association (ACC/AHA) developed a Heart Risk Calculator for doctors to predict the risk

BodyPartRecognition.pdf, June 1, 2011, accessed November 27, 2016
[13] Cardiovascular Diseases (CVDs), World Health Organization, May 2017, http://www.who.int/mediacentre/factsheets/fs317/en/

of heart attack in the next ten years.[14] The calculator takes into account eight risk factors: age, gender, race, both total and bad cholesterol levels, blood pressure, weight, tobacco use, and the presence of diabetes. While the Heart Risk Calculator considers only eight risk factors, cardiovascular disease results from *many* factors such as a family history of heart disease and arthritis. Researchers at Nottingham University in the UK led by Stephen F. Weng conducted research to determine if supervised machine learning could make better predictions of heart attack than the Heart Risk Calculator recommended by the ACC/AHA. In the paper titled, "Can Machine Learning Improve Cardiovascular Risk Prediction Using Routine Clinical Data?" the authors used several types of supervised machine learning, including random forest, to build a prediction system.[15] With 374,000 patient records from the UK Health System, the researchers took a set of records of people who had no heart attacks recorded in 2005 and let the machine learn from the records what types of symptoms and risk factors contributed to heart attacks in the years following 2005. Once the computers were trained on patient records, new records were fed to the machine to see if it could predict heart attacks. Interestingly, the researchers found that the supervised machine learning algorithm produced significantly better predictions than the Heart Risk

---

[14] Heart Risk Calculator, American College of Cardiology/American Heart Association, 2013, http://www.who.int/mediacentre/factsheets/fs317/en/

[15] Stephen F. Weng , Jenna Reps, Joe Kai, Jonathan M. Garibaldi, and Nadeem Qureshi, "Can machine-learning improve cardiovascular risk prediction using routine clinical data?," PLOS|one, April 4, 2017, https://doi.org/10.1371/journal.pone.0174944, accessed online March 24, 2018

Calculator. For example, the ACC/AHA accurately predicted a heart attack 72.8% of the time while the supervised machine learning algorithms ranged from 74.5-76.4% accurate. With millions of people affected by cardiovascular disease, such an improvement in prediction would translate to many lives saved with the application of proper intervention. Moreover, the machine learning found patterns that the doctors did not. The Heart Risk Calculator considered diabetes and high blood pressure to be significant contributors to the risk of heart attack, but random forest did not rank either these in the top eight risk factors. Instead, random forest found that body mass index and poverty made for stronger predictors of a heart attack. The sometimes-subtle relationships between different health factors may escape a doctor's eye or clinical bias, but supervised machine learning can consume vast amounts of information (without bias) and reveal relationships that experts may look over.

Supervised machine learning helps with classification and prediction. Another strength of supervised machine learning comes from its unbiased view of the information from which it learns. Sometimes a variable that experts ignore may be far more important than initially thought. Once random forest has learned the best decision tree, it takes new data, and it predicts an outcome like a house price or risk of a heart attack using the patterns it learned earlier. The analysis of customer churn in the telecom industry compared the results obtained with machine learning with more traditional analytics, and the Azure team found that their model outperformed more conventional methods in identifying the highest flight risk customers. Utilizing machine learning tools to understand customer churn addresses a fundamental problem faced by most businesses. Think about your business and what types of data you may have or could efficiently collect about your customers to predict churn or any other kind of question that asks will it be A or B?

Unsupervised Machine Learning

"Reports that say that something hasn't happened are always interesting to me, because as we know, there are known knowns; there are things we know we know. We also know there are known unknowns; that is to say we know there are some things we do not know. But there are also unknown unknowns—the ones we don't know we don't know. And if one looks throughout the history of our country and other free countries, it is the latter category that tends to be the difficult one."

—US Defense Secretary Donald Rumsfeld [16]

The narrow artificial intelligence known as unsupervised machine learning lets the computer learn all on its own without help from people in contrast to supervised machine learning which needs people or another machine to tell the computer what it is learning. Previously, we were looking at the strengths of supervised machine learning in classification, ranking, and prediction—using patterns and categories you know

---

[16] http://archive.defense.gov/Transcripts/Transcript. aspx?TranscriptID=2636, DoD News Briefing - Secretary Rumsfeld and Gen. Myers, Presenter: Secretary of Defense Donald H. Rumsfeld' February 12, 2002 11:30 AM EDT. Transcript of accessed online December 11, 2016

to train the system to work with new information. In some cases, though, you may not be looking for patterns that you know. You may need to learn what you do not know. Donald Rumsfeld's famous "unknown unknowns" reminds us that we do not always know what we are looking for, and unsupervised machine learning excels at finding anomalies, weird things, strange things— things that unexpectedly do not fit. As a reminder, unsupervised machine learning identifies patterns by itself without training information like the previously mentioned supervised machine learning. Supervised machine learning uses known data to find more known information, but in unsupervised machine learning, the thing you find may occur so rarely that you cannot train for it. In other words, there is no teacher, no parent, no authority telling the machine what it knows. Unsupervised machine learning excels at clustering and associating data. The difference between clustering and associating is that clustering finds groups, and associating determines rules in the data. For example, clustering may decide that one group of customers that come to a mall to buy books, and association will learn that when a person buys a book on yoga, they also buy a book on spirituality. Unsupervised machine learning, in other words, finds patterns. The unsupervised machine learning opportunities will help you find unusual and unexpected patterns that are valuable for security purposes, revealing new opportunities for investing money, or finding new customers. Unsupervised machine learning helps answer questions like:

◊ Is this a fraudulent credit card transaction?

◊ Does this employee usually download this type of information?

◊ Is the current temperature reading in this jet engine typical at this altitude?

◊ Do my service trucks ever stop for this long in Nevada?

Finding the anomalous, the weird, or the unusual has clear implications for security, and security has benefited dramatically from anomaly detection. Credit card companies routinely sift through millions of transactions, identify unusual transactions, and flag them for fraud investigation. With identity theft on the rise in combination with more extensive and more diverse data becoming available every day, security analysts desperately need computational help to identify data theft and to better predict when attacks may happen. Unsupervised machine learning contributes significantly to cybersecurity and helps determine the unusual or *asynchronous* access patterns of a data security breach. Symantec reported in a 2016 white paper an estimate that cybercrime cost the world economy $575 billion that year.[17] Improvements in catching cyber thieves will remain big business for years to come. As history has shown, every time the security gets better, eventually so do the thieves.

Reinforcement Machine Learning

Reinforcement machine learning helps machines figure out the best behavior to get the best results. Reinforcement machine learning learns much the same way that humans and animals learn—by repeating the

---

[17] "Internet Security Threat Report," Symantec, Vol. 21 April 2016

things that are good and not doing the things that hurt us or for which we are punished. In reinforcement machine learning, the computer is not told what to do, but it experiments with different actions and receives positive feedback for the right response or negative feedback for a wrong decision. People learn with the help of others or on their own what are right actions and what are wrong actions. Pain is a potent reinforcement teacher. When I was very young, I remember being in my grandma's kitchen staring at the steam coming from the tea kettle on the stove and hearing the loud whistle. My grandmother told me not to touch the steam, but it looked just like clouds, and I didn't listen. I stuck my hand into the steam, and it instantly burned. It was immediate and painful. I never did that again. This type of learning is called *cause-effect* learning because we directly experience the results of our actions.

For reinforcement learning, the same process holds for computers that use reinforcement learning. Computers can get feedback from a user, another computer or other sources such as visual, mechanical, audio, chemical, etc. Success gives positive feedback. For example, picking stocks and being rewarded when the pick is right—the stock price goes up—and getting a demerit for picking a poorly performing stock. When computer scientists write computer code for reinforcement learning, they include a part of the program that compares the outcome of each action such as comparing the score in a game to an opponent's score. When the score for the computer is higher than the opponent's score, the computer saves the actions it took to win as a positive and will use them in future games. In fact, reinforcement learning works well in playing games, because the high score serves as a positive reinforcement. Researchers at Google DeepMind in London published a letter in the science journal *Nature* describing how they developed a reinforcement learning algorithm that learned to play Atari 2600 games with only the input of the screen pixels and the score. The algorithm called a Deep-Q Network was able to master 46 computer games to the level of professional game testers with no initial input.[18]

Reinforcement learning applies a combination of exploration, trying new things, and exploitation, wanting the best outcome.[19] What makes reinforcement learning so compelling is that a computer can work continuously and at a very high speed to play a game over and over again until it figures out how to win. Reinforcement machine learning has applications in many areas of the public and private sectors. For example, the telecom industry, which routes billions of phone calls and text messages every day, must maintain a vast network of computers, switches, and transmitters. Keeping the system up and running requires constant maintenance and attention. Traditionally, network failures would arise, and a person would have to figure out how to fix it. But with reinforcement machine learning, computers can not only figure out how to make the repairs but even predict problems before they happen and take appropriate actions to prevent a failure.[20]

A powerful extension of reinforcement learning applies to robotics. A teachable robot that can learn

---

[18] V. Mnih, et al., "Human-level control through deep reinforcement learning," *Nature*, Vol. 518, p 529, Feb. 26, 2015
[19] Faizan Shaikh, "Simple Beginner's guide to Reinforcement Learning & its implementation," *Analytics Vidhya*, January 19, 2017
[20] Ajit Jaokar, "The applications of Artificial Intelligence (AI) in the Telecoms industry," *Data Science Central*, April 16, 2017, accessed

different skills, like a bright worker, would quickly become a valuable asset. Trainable robot arms became commercially available not long ago. Instead of programming the moving and grasping arm (remember the peanut butter and jelly sandwich example of programming in Chapter 1?), the robot can learn by doing. Now, this type of learning becomes more valuable as the robot is required to do more complex operations in more unpredictable environments. In many ways, the robot will need to learn just as people learn. Think of learning to hit a tennis ball over the net. The first time you tried, you may have missed the ball altogether, hit it too softly to go over the net, or, like me, hit it so hard it flew out of the court. I felt some negative feedback from that experience—having to exit the court and find the ball in an adjoining house's yard. So, next time, I hit the ball softer. Now I did not note my exact arm speed or the speed of the tennis ball; I just hit it a bit more softly, and the ball hit the net. The next time, I hit the ball a little harder, and this time, the ball sailed nicely over the net. It was a gratifying moment. I have hit the ball thousands of times since then and have become proficient at tennis. Successful shots give a feeling of accomplishment, a reward and bad shots get a negative response. In the same manner, robots using reinforcement machine learning become proficient at tasks they were not explicitly programmed to do.

A significant use of reinforcement learning applies to the field of robotics, which gets covered in more detail below. Robots learn how to manipulate objects and sort them for applications in manufacturing, supply chain management, and product delivery. For example, Amazon makes extensive use of reinforcement learning robots in sorting their massive warehouse inventories.

Deep Learning

In supervised machine learning, a computer receives data inputs such as pictures. At a basic level, the computer *sees* the picture as a collection of pixel values with no meaning or context. Each pixel makes up the tiny area of a computer or television screen that displays a color at a certain intensity level. All the pixels lit up with the appropriate colors across a screen will produce a picture. Whether the picture is static or moving, the computer *sees* a picture as a series of pixels, and each pixel has its mixture of red, green, and blue (RGB) light that represents every pixel in the picture. Remember the computer in your TV or camera will faithfully reproduce an image, but it does not know what it is displaying. It is merely following orders on what color each pixel should show—how all the tiny lights in the screen should glow to make a picture. So how can a computer make the jump from blindly following orders to displaying an image to learning what is in the image? The computer will use what is called *deep learning* to decipher what is in the image, extracting essential features like faces, objects (such as cars, trees, and chairs), and even tell if the image is indoors or outdoors by a river or in the mountains.

Deep learning uses *neural networks*. Neural networks are not like real nerves found in human brains, but the concepts of data processing did come from discoveries in neuroscience back in the 1940s. The theory says that data can be processed in a machine

The Rise of Artificial Intelligence

much like in a brain if we model how nerves work together. Nerves receive input information, and when enough of that information comes in, the nerve will fire, sending a signal to the next nerve. Human sensing of heat illustrates the point quite clearly. In the case of heat, recall putting your finger near a candle flame. You will feel the increasing heat when you move your finger closer to a candle flame. The heat-sensing nerves in your finger receive input from the burning candle. When your finger is far away, there is not enough heat to make the nerve fire and send a signal to the brain, but once your finger gets close enough to the flame, a nerve will reach a threshold and say, "This is hot," and send a message to the next nerve. When the nerve fires, it will communicate with the next nerve down the line. If enough heat-sensing nerves are saying, "It's hot!" the next nerve in the chain that connects your finger to your brain will fire, and your brain will say, "That's hot!" Nerves process the input from the outside world, and the brain detects the presence of heat. Now if you add the visual picture of a candle, and the brain joins candle flame and heat, you learn that candle fire is hot. In essence, using the same way that nerves take in information and decide whether or not to send the signal to the brain, machines can contain neural networks. Neural networks break down input information and sort the information through different filters. After sorting out details from thousands, even millions of inputs, the computer learns patterns that will help it figure out what it is looking at when it sees something new. For decades, neural networks in machines were not possible because the computers were too slow, did not have enough memory, and there just was not enough data to train the machines. All of that changed in the past ten years with the advent of faster computers, cheaper and more abundant memory, and the arrival of big data.

As mentioned earlier, facial recognition has evolved by leaps and bounds. The researchers at Baidu claim that their system for facial recognition has a 99.7% accuracy rate, which is better than a human.[21] Part of what makes their facial recognition so compelling is the use of what are called *convolutional neural networks*.[22] Convolutional neural networks constitute a type of deep learning used extensively in facial recognition and object identification that works in a manner much like a brain does. The convolutional neural network in the computer first pulls simple things from an image such as bits of color, curved lines, shapes, and textures, and as the image goes through the network, larger features are then pulled such as hair, ears, and eyes. Ultimately, after the deep network has processed thousands or even millions of faces, it has learned to recognize faces and can do so very quickly. Unlike with supervised machine learning the deep learning artificial intelligence will discover on its own what a face is and how to recognize a face it has not seen before.

The victory of Google DeepMind's AlphaGo over the world Go champion Lee Se-dol marked a significant advancement in artificial intelligence and showcased the power of deep learning. The game of Go has been played for over 2,500 years and was considered

---

[21] Jingtuo Liu, Yafeng Deng, Tao Bai, Zhengping Wei, Chang Huang, "Targeting Ultimate Accuracy: Face Recognition via Deep Embedding," arXiv:1506.07310 [cs.CV],2015, accessed online on September 26, 2017
[22] Sean Captain, "Baidu Says Its New Face Recognition Tech Is Better Than Humans at Checking IDs," *Fast Company*, November 17, 2016. Accessed on line September 30, 2017

in ancient times one of the four essential arts of the Chinese scholar (the others being music, calligraphy, and painting.) Go is a board game for two players with a 19 x 19 grid and black and white stones, black for one player and white for the other. Players take turns placing their stones on the grid. The goal of Go is to enclose more territory than your opponent. Due to the large grid, there are billions and billions of possible moves. Go is considered more complicated than chess and has been viewed as the one game at which a computer could not beat a human. Enter AlphaGo. Google DeepMind created AlphaGo, which used artificial intelligence to master the game of Go. Instead of writing a complex set of rules for the computer to use, the computer scientists let AlphaGo learn to master the game on its own. By playing against itself millions of times and learning each time how to improve its game, AlphaGo worked up the ladder to become a champion Go player. In a landmark achievement for artificial intelligence on March of 2016, AlphaGo defeated the world's greatest Go player, Lee Se-dol, in 4 out of 5 games played. What makes the victory so interesting is that AlphaGo by playing itself created its own strategy that was different from expected human moves. Fan Hui, an international Go champion, remarked when watching AlphaGo defeat Lee Se-dol, "It's not a human move. I've never seen a human play this move. So beautiful!"[23] One of the English language commentators covering the man v. machine match noted, "That's a very surprising move. I thought it was a mistake." After the matches were over, Lee Se-dol conceded that he never felt in control

---

[23] Cade Metz, "The Sadness and Beauty of Watching Google's AI Play GO," *Wired*, March 11, 2016, accessed online June 2017

of the games. AlphaGo really shocked him with its ingenuity.[24]

Artificial General Intelligence

To this day, all of the artificial intelligence that has been deployed in the public and private sectors has been narrow artificial intelligence. The performance has been spectacular, but researchers have not developed artificial general intelligence or strong artificial intelligence. Artificial general intelligence goes beyond the machine learning and deep learning that we have been looking at so far. Artificial general intelligence would be the case of a fully independent, *conscious* entity capable of surviving and thriving on its own, an entity with its own will and motivations to go out into the world and live among us. Artificial general intelligence contains a concept referred to in popular culture as the *singularity*. The singularity is a point in time when the intelligence of our machines surpasses human intelligence.[25] Ray Kurzweil and other futurists have been saying, "The singularity is near!" for some time, meaning at some point in the next 20-100 years. But what will happen at the singularity? Will humans fall back a peg on the evolutionary scale? Will people and machines become more intertwined? We will return to these questions throughout the book and look at the issues facing us now culturally, ethically, economically, and personally. Before looking to the future though, we will begin by

---

[24] Cade Metz, "How Google's AI Viewed the Move No Human Could Understand," *Wired*, March 14, 2016 accessed online June, 2017
[25] Ray Kurzweil, *The Singularity Is Near: When Humans Transcend Biology*, Penguin Group, 2005

looking at how we are already living with and among AI, machine learning, and robotics.

Chapter 3
Applications of Artificial Intelligence

Artificial Intelligence, or AI, refers to computers that mimic or exceed human capabilities such as decision making, creativity, problem-solving, pattern recognition, purpose, and perhaps someday, consciousness. Artificial intelligence needs to consume information to properly function and grow. The information may come from sensing the world through vision, touch, hearing, and smell. Moreover, data may come to a computer through spoken or written language or symbols like numbers. The learning aspect of artificial intelligence is human-like, as it will learn from its mistakes, make guesses with incomplete information like a person does, and even develop the ability to listen and speak. Artificial intelligence applies to different tasks such as language translation, the identification of a person seen on a security camera, or the careful touch of a robotic surgeon. For this book, we will focus on some of the significant areas of development in AI with a more in-depth look into machine learning and demonstrate how this is co-evolving with robotics. The central regions of focus will be perception, knowledge, natural language processing, learning, planning, social intelligence, creativity, and a brief look at artificial general intelligence. In this chapter, we will look at powerful applications of artificial intelligence that demonstrate how artificial intelligence perceives the world, deals with language, planning, creativity, and decision making.

Perception

Perception, the ability to see, touch, smell, in short, the ability to know what and who is out there makes us who we are and allows us to navigate the world around us. Artificial intelligence takes in information in different forms such as text, sound, or sight and through learning by trial and error can process the input and return an output such as recognizing faces, spoken words, or decide what action to take. For example, artificial intelligence helps a driverless car to avoid a pedestrian crossing the road. Artificial intelligence uses inputs like the five senses of sight, sound, touch, taste, and smell. Remember that artificial intelligence acts more like a human because it will learn from its mistakes, make guesses with incomplete information, and even develop the ability to listen and speak

Intelligence requires perception. Think for a minute that a computer as it comes off the assembly line has no understanding of the world around it. The computer is blank. Without complex machine learning, software programs, and human-made sensors, the computer could not possibly understand the world around it. Perception remains a central focus of artificial intelligence, and computer vision has been the focus of much research and exceptional achievement over the past 20 years.

To understand the achievement of computer vision, think about a digital picture on a computer, digital camera, or smartphone. Just hooking up a digital camera to the computer does not mean the computer can *see* what is in the pictures streaming into its memory. All that is flowing into its memory is a vast field of numbers. The computer represents colors as a series of numbers that reflect the intensity of red, green, and blue for each pixel or dot in a picture that the camera captures—combinations of red, green, and blue that cover all the possible colors of the rainbow. The computer can faithfully save the image file to memory or display the picture on the screen, but it does not know what is in the picture. At one layer, it is just numbers—information with no meaning. The little computer in your camera can perform transformations of the data with lightning speed, but the vast field of numbers does not translate into a picture of your friend to the computer. Unless the computer is given machine learning algorithms to determine what is a human face or other objects, it will have no sense or interpretation of what the digital picture contains.

Computer scientists have developed more and more capable object recognition algorithms in recent years, and artificial intelligence has been instrumental in this progress. In a classical programming way, one could imagine that a computer would not have to be *intelligent* if there were a way to store labeled pictures of everything in the world in a giant database. The computer could just match any picture with one in its database, and everything would be clear. For example, a picture of President Eisenhower could be labeled in a vast database, and if the computer were to see the picture of Eisenhower, it would make an exact identification by finding a match in the database. But there is no possible way to have a picture of everything perfectly labeled as to what it is. Everything is moving and changing and looks different from different angles. But with artificial intelligence, computers do not have to match everything to known examples. The machines can learn what a human face looks like and use this learning to identify human faces in pictures it has never seen.

How do computers, even the tiny computers in your digital camera or your smartphone, learn what a face is and find it in a picture? Instead of having a description of all possible faces, the computer program will examine pictures of faces, and it will make measurements of many features such as eye shape, distance between the eyes, and the placement of eyebrows above the eyes which are above the nostrils. Given this training set, the computer will make up some rules that apply to faces. The computer can then see a new face using the rules that it wrote for itself and determine if it is a human face or not. Computer vision is improving continuously, but this is just the beginning.

Try to remember the last time you looked at an early childhood reading book. Early reading books mostly have big pictures and some action-filled text. The classic early reader *Go, Dog. Go!* by P.D. Eastman combines great illustrations and rhyming language to help early readers connect the pictures and words.[1] The books may look simple, but the task of connecting pictures to things in the real world represents a very complex activity. Moreover, the words describe the action of the static images adding another considerable complexity to the process of learning to read. Not only does the reader need to recognize two-dimensional drawings of a dog on skis and bikes, but must also be able to read the text that describes all the different types of transportation taken by the dogs.

Storing information in a book with pictures and words to help children read represents a concept called *knowledge representation*. Similarly, for artificial intelligence, knowledge representation involves the storing of information regarding the real world that a computer can use to perform tasks, make decisions, and make predictions. In a powerful program developed by the US Defense Advanced Research Projects Agency (DARPA) described as a search engine on steroids, vast amounts of public data get processed and then analyzed to counter the terrible crime of human trafficking.[2] The program DARPA developed called MEMEX collects and examines a wide variety of information from ads placed in local newspapers and online, names, pictures, internet links, and more. MEMEX does not merely return links or top hits. Instead, it goes many steps further and analyzes the contents of images, the text in advertisements, and the geolocation of pictures. For example, MEMEX does not simply do facial recognition but scans the background of photos

[1] P.D. Eastman, *Go, Dog. Go!*, Random House Children's Books, 1961
[2] *Wade Shen, "Memex," Defense Advanced Research Projects Agency, https://www.darpa.mil/program/memex, Accessed online November 2016*

to identify places such as certain hotel rooms or parks. Human traffickers may use the same location over and over again to build a portfolio of their sex workers used in advertising and promotions. Such clues help MEMEX learn of the location and size of trafficking rings and provide clues to who is running the ring. MEMEX not only gathers data from the web regularly used by businesses and average users, but it also delves into the Dark Web of anonymous servers popular with more illegal activity and services such as the now-defunct Silk Road drug market. Also, names used in ads and the phrasing of the ads form associations that MEMEX identifies. The result of this data mining plus analysis builds visual networks connecting locations and people that law enforcement agencies across the country use. Sometimes the trafficker will use the same advertisement in a new city or town or post the same name or phone number. Machine learning will point out this pattern and help law enforcement rescue trafficked women and children from the sex industry. MEMEX embodies supervised machine learning, and the example powerfully demonstrates the potential to construct useful information that in disconnected bits of data alone would not be useful. The streams of data from pictures and advisements together with analysis reveal criminal networks and give an edge to law enforcement.

Knowledge representation in AI encompasses the vast, almost uncountable, associations of information you take for granted in your mind that allow you to make decisions. Take for example the simple case of looking at an object and guessing if it is heavy or not. First, you must know your strength, and secondly, you must use knowledge gained through experience and reason about what is heavy. Think of a series of rocks lined up from a pebble to a medium-sized rock to a boulder. Just looking at the stones, you could probably tell which ones you could pick up and throw, which ones you could maybe pick up and carry, and which ones you could not budge even if you pushed on it. The complex associations of recognizing rocks and your experience with them in life combine into a knowledge representation that would help you make a decision about which ones you could move. Your estimation capabilities would come into play if a rock slide were blocking a road, and someone were to ask you to help clear the road. You would not waste your time trying to move a big boulder without help from others with tools or machines. You would determine which rocks you could move and begin there. The rock example should make you appreciate the complexity of representing information for a machine to understand: identify some object like a rock, determine how heavy a rock is in relation to one's strength, and decide to move the rock.

Language Processing

A room full of people talking makes a deafening sound because people like to communicate and do it often with words. We use spoken and written words all the time to share our thoughts and emotions, and it makes sense that successful artificial intelligence must be able to understand and communicate with us using human language.   Many science fiction stories feature people talking directly to an ever-present computer. The television series *Star Trek* featured an omnipresent computer that the characters often called upon to provide information or to help solve problems.

Computer science uses natural language processing to help computers understand spoken and written human language. Matt Kiser describes it well in "Introduction to Natural Language Processing."

"NLP [natural language processing] is characterized as a hard problem in computer science. Human language is rarely precise, or plainly spoken. To understand human language is to understand not only the words, but the concepts and how they're linked together to create meaning. Despite language being one of the easiest things for humans to learn, the ambiguity of language is what makes natural language processing a difficult problem for computers to master."[3]

To understand the challenge a computer faces in natural language processing, think of a concept and the information that gives that concept meaning. Take the word *fly*. If you hear that word alone without any context, it is hard to know if it is referencing a house *fly*, *to fly* like an airplane, or the slang word of *fly* meaning cool. It becomes even more confusing for a machine considering the sentence, "They watched the fly fly and thought it was fly." Natural language processing helps computers make sense of situations like the above sentence. For a computer to understand the sentence, it either needs a series of structures and rules that support it to determine that the verb in the sentence refers to that action of flying and the object is the insect, or it has to learn just as people do by repetition and example. Previous efforts in computer science focused on the rules of how language works and met with limited success in genuinely understanding

---

[3] Matt Kiser, "Introduction to Natural Language Processing (NLP) 2016," *Algorithmia*, August 11, 2016, accessed online October 8, 2017

human language.

Successful natural language processing benefits immensely from artificial intelligence, especially deep learning. Using artificial intelligence, computers learn languages and meaning instead of following rules leading to significant improvements in natural language processing such as:

◊ Speech recognition

◊ Sentiment analysis

◊ Language translation

On February 14-15, 2011, the world watched IBM's supercomputer, Watson, win the question answering game show *Jeopardy!* by soundly defeating two of the show's greatest champions, Brad Rutter and Ken Jennings. In the defeat, Watson earned three times more money than the nearest human competitor.[4] People watched in amazement as the computer would listen to the questions being read and answer them in its not exactly human voice. Watson was built as a question-answering machine specifically to compete on *Jeopardy!*. Behind the voice and logo visible to the public on the game show, IBM computer scientists had developed a very sophisticated and powerful computer. Part of Watson's computing power came from its ability to recognize human speech so that it could break down the questions given in English in order for Watson to provide an answer. Watson used machine learning to perform speech recognition in real time. Moreover, for Watson to prepare for *Jeopardy!*,

---

[4] "Computer crushes the competition on 'Jeopardy!'" AP News, 2011, accessed on the internet on September 12, 2017

its developers gave it 200 million documents, including dictionaries, encyclopedias, books, and more to help it learn.[5] Watson analyzed all the millions of records with natural language processing and used the text to formulate its answers to questions with remarkable speed and accuracy. Watson stands as a monumental step forward in artificial intelligence, including natural language processing. Things continue to improve, and many applications for speech recognition surround us today from the automated help over the phone to the proliferating intelligent assistants such as Apple's Siri, Microsoft's Cortana, or Amazon's Alexa.

If only we knew what people were thinking. Understanding the sentiment of a document or an entire population provides precious information for businesses, governments, and individuals. Natural language processing gives analysts powerful and rapid access to how people are feeling about a topic more rapidly and accurately than ever before. Sentiment analysis is the process of determining from text and social media sharing, whether people are feeling positive, neutral, or negative about a subject such as an event, advertisement, or movie. Businesses use sentiment analysis to determine the success of an advertising campaign or reaction to their brand after some bad news. Natural language processing provides the platform to analyze text for social sentiment. The social media company Twitter provides a platform that rapidly shares short statements, or *tweets*, to millions of Twitter followers in seconds. Analysts follow the

---

[5] Katherine Noyes, "It's (not) elementary: How Watson works," *PC World IDG News Service*, October 7, 2016, accessed online October 9, 2017

reaction to them to get a real-time understanding of trends in sentiment.

Many tools have been developed to do Twitter sentiment analysis.[6] United Airlines was the subject of a storm of negative tweets after a video depicting officials dragging one of their passengers, bloody from their plane. The video rapidly spread across the internet. An analysis of tweets following the event showed a massive spike in negative and sarcastic tweets against United Airlines.[7] As time passes and United tries to rebuild its brand and reputation, continuing sentiment analysis provides a good real-time measure of the success or failure the marketing efforts.

Star Trek fans have marveled at the concept used in the science fiction series of a universal translator that allows free dialog between humans and alien species. Anyone who has struggled to learn a foreign language has wished for a real-time voice translator for travel, business expansion into foreign countries, and meeting new people who speak a foreign language. The future is here, and one of the most potent technologies developing out of natural language processing is the application of real-time language translation. Using artificial intelligence in the form of deep learning recurrent neural networks, Google among others have vastly improved real-time language translation.[8] Google Translate supports language-to-language translation of

---

[6] Daniel Harris, "The Best Free Tools for Twitter Sentiment Analysis," SoftwareAdvice.com, accessed online November 21, 2017
[7] Vicky Qian, "Step-By-Step Twitter Sentiment Analysis: Visualizing United Airlines' PR Crisis," iPULLRANK blog, April 26, 2017,
[8] Mike Schuster, Melvin Johnson, and Nikhil Thorat, "Zero-Shot Translation with Google's Multilingual Neural Machine Translation System," Google Research Blog, November 22, 2016, accessed online October 8, 2017

over 100 different languages and processes over 140 billion words per day. In 2016, Google announced the move from Google Translate to Google Neural Machine Translation (GNMT). GNMT deploys a much more powerful artificial intelligence that not only translates languages that it has learned but can translate even between languages that it has not yet learned. For example, once it learned to translate between English and Japanese and English and Korean, it was able to translate between Japanese and Korean. Technology now exists that can decode the human voice and convert that sound into words.

With natural language processing, the computer can translate the spoken word to another language and use a synthetic voice to speak to another person. Recently, some international conferences employed real-time translation for the attendees. One company called TYWI, or Translate Your World Incorporated, based in Cupertino, CA serves as an aggregator of all the best in automated language translation from Microsoft, Google, Apple, Baidu, and other technologies for real-time translation for use in presentations and via video conferencing.[9] On November 10, 2017, Google announced a product called Pixel Buds that the wearer could use for near real-time translation.[10] According to the press, Pixel Buds are wireless earphones that need a Pixel (Google's interactive assistant), an Android phone, and an internet connection to provide translation simultaneously of 40 languages with only a 1-2 second

---

[9] Translate Your World, http://www.translateyourworld.com/en/about/
[10] Lucas Matney, "Google's Airpods competitor do real-time language translation," TechCrunch.com, Oct 4, 2017, accessed online December 1, 2017

delay for translation. No one yet is claiming perfect translations with all the nuance of a native tongue, but the time is upon us that computers with artificial intelligence can simulate having a personal translator with you at all times.

Planning

Planning a fun vacation requires the anticipation of many variables from the right time to go, who can come, where to stay, how to get there, what to do when there, and more. Planning, in general, requires the anticipation of a series of events and the actions to take at each stage to achieve a particular goal. Computer-assisted planning has been studied for many years, and for very controlled environments where all the variables are defined artificial intelligence is not needed. Enterprise Resource Planning, or ERP, is a good example.[11] Developed in the 1990s, ERP developers required complex models of how companies worked and wrote complex planning software to help with overall enterprise planning from incoming materials to outgoing products. As discussed earlier, stable, predictable environments lend themselves to classical computer programming with all the rules and defined to do planning. However, many situations are more dynamic. To continue with the vacation planning analogy, planning a vacation to a resort or theme park provides a finite set of options—which hotels to stay in, what rides to take, which restaurants to eat at, and

---

[11] F. Robert Jacobs, F.C. 'Ted' Weston Jr, "Enterprise resource planning (ERP)—A brief history," *Journal of Operations Management* 25, 357–363, 2007, accessed online October 9, 2017

more. For a vacation like that, a traditional planning tool would work, but what about a vacation with more variables like a backpack tour across India or Europe? Sometimes, one could stay with friends, camp out, or get a hotel room if the budget permitted. The situation is more dynamic, and the planning is more difficult. Artificial intelligence lends itself to planning in more dynamic conditions.

The Jet Propulsion Lab (JPL) at Cal Tech and NASA tackled very complex, automated planning scenarios using artificial intelligence. NASA faces elaborate planning and execution for remote space probes and distant planetary rovers. "The basic problem is to develop a sequence of commands for a system that achieves the objectives of the user of that system. The user (e.g., scientist) has some high-level objectives, or goals. Typically, the system (e.g., spacecraft) has a low-level command interface. Therefore, the problem becomes translating the high-level goals into a valid sequence of low-level commands. At JPL, some of the primary systems that require commanding are: deep-space probes, planetary rovers, and deep-space communication antennae."[12] To solve these complex problems, NASA turned to artificial intelligence and developed a system called ASPEN, Automated Scheduling and Planning ENvironment. ASPEN allows planning and scheduling for remote actions and rapid rescheduling if a problem is encountered in dynamic environments millions of miles from earth. An extension of artificial intelligence use at NASA is the Continuous Activity Scheduling Planning

---

[12] ASPEN, https://aspen.jpl.nasa.gov/

Execution and Replanning, or CASPER project.[13] The ground breaking and award-winning CASPER project enables autonomous spacecraft and planetary rovers to travel and work on their own without being driven continuously remotely. With CASPER, spacecraft and rovers dynamically adapt to changing situations while maintaining mission goals and success. Planning and scheduling capabilities found in ASPEN and CASPER have many terrestrial applications as well, and growth in this area of artificial intelligence should continue to penetrate both the public and private sectors.

Creativity and Artificial Intelligence

"As technology increases the role of creativity will become more not less important. We must still confront that part of the problem-solving task which the machines cannot emulate, the ineffable part—creativity."[14]

—Steven A. Schwartz

In many respects, creativity is man's most potent attribute. Creativity—the powerful ability to combine ideas in unique ways to solve problems—has elevated man out of subsistence to exceptional levels of culture, art, science, and technology. What is creativity? We need to look at this question to position ourselves to best protect ourselves and profit from the rapidly advancing world of artificial intelligence, machine

---

[13] AI@JPL, CASPER, https://www-aig.jpl.nasa.gov/public/projects/casper/
[14] Stephen A Schwartz, "Creativity, Intuition and Innovation," *Subtle Energies*, Vol. 1, No. 2, 1991

learning, and robotics. Scholars, psychologists, social scientists, and even computer scientists vary on the definition of creativity, but the meaning refers to the development of novel ideas or things that did not exist before which solve a problem. Another way to describe creativity is the moment of insight like a flash, or an instant knowing, and the solution to your problem appears. In his paper, "Creativity, Intuition and Innovation," Stephen Schwartz, author and researcher, suggested that these three elements of the human mind have received attention by different disciplines. The disciplines include psychology, neurology, sociology, physics, and art, but the linkage between them has not been looked in any satisfactory way.[15] He aptly quotes Einstein as saying, "I believe in intuition and inspiration [...] Imagination is more important than knowledge. For knowledge is limited, whereas imagination embraces the whole world. Giving birth to evolution." The impact of creativity may range from incremental advances by a scientist finding a way to improve a chemical reaction by a few percentage points to world-changing creativity such as the creation of a new type of mathematics.

Back in 1687, Sir Isaac Newton, the English physicist and mathematician, published *Philosophiae Naturalis Principia Mathematica*, which gave birth to calculus and is considered one of the most significant books ever written in science. Before Principia, there were substantial branches of mathematics such as geometry and algebra but no calculus. Newton as a physicist struggled with the available mathematics of

---

[15] ibid

his time to explain what he observed in the natural world. When objects fell to earth, they moved faster and faster; however, he could not describe it mathematically. Without the proper mathematics, he could not predict how quickly an apple would fall from a tree. Geometry didn't help. Algebra didn't help. So, he invented an entirely new branch of mathematics that could better describe the changing natural world around him. Calculus is the math of changes, and Newton created it to help him solve some significant problems. Just look around, and you will see the evidence of human creativity everywhere from cars to houses, to airplanes, to artwork, to even computers. The nature of human creativity has sparked much debate and interest in many realms of research, including computer science. In fact, the discipline of computer science has an area specifically surrounding creativity called *Computational Creativity*. Computational Creativity focuses on understanding human creativity and tries to model it in computers.

A genuinely compelling question is can a computer with artificial intelligence be creative? Some thinkers such as Bernd Schmitt of Columbia University commented recently in an interview for the *MIT Sloan Management Review* that "Creativity is, basically, knowing a lot about a domain and then making sense of an unexpected event and adjusting to it by devising new solutions. Computers can certainly do that."[16] While this book is not the forum for debating on the true nature of creativity, other thinkers and philosophers

---

[16] Frieda Klotz, "Are You Ready for Robot Colleagues?" an interview with Bernd Schmitt, *MIT Sloan Management Review*, July 6, 2016, accessed online July 27, 2017

will argue that from a scientific and engineering perspective, creativity involves the ability to materialize novel ideas for solving problems. Can we say that a computer with access to vast stores of knowledge and the mandate to create novel connections is not creative? Part of the creative process is the evaluation of novel ideas. It may be novel to combine grass clippings and rocks for breakfast, but it is not a good idea—it makes no sense. More importantly, solving problems still has a science and math bias, and art, the most powerful form of human creative expression, does not lend itself to artificial intelligence. Art is not made to solve problems, rather, it is a call to self-expression. We can say that a computer does not make art to express itself. A computer will not drop its facial recognition task to write a poem about a particularly enchanting face it came across. Humans will program computers to make *art*, but the computer is simply the medium like a camera, not the artist like the photographer.

Today, we allocate some writing tasks to machines. Take, for example, the use of machine-generated sports reporting, so-called robo-journalism. A product called Wordsmith is made by Automated Insights located in Durham, North Carolina. The company describes their product as "… a natural language generation (NLG) platform that turns data into insightful narratives."[17] By taking data from a sporting contest, Wordsmith will write a news story about the game. Beyond the regular attributes of the game such as reporting the winner and loser, Wordsmith selects unique elements of the contest to highlight the game using statistics and supervised

---

[17] "Automated Insights," Wordsmith, https://automatedinsights.com/wordsmith, accessed online October 9, 2017

machine learning then generates a story fit for print. Wordsmith powers more than just sports reports. It converts data into stories for financial reports, hotel descriptions, weather reports, and more. In fact, the company claims to generate more than 1.5 billion pieces of content per year for diverse customers from the Associated Press to Microsoft and more. Robo-journalists do not convince everyone. Emily Reynolds reported in "Wordsmith's 'Robot Journalist' Has Been Unleashed" that NPR (National Public Radio) listeners found the article written by a human "richer and more engaging," than the machine generated article.[18] Robo-journalism is a catch-all term for machine writing, and as far as creativity goes, it may not be on par with human writing in all cases, but it is performing a task for hundreds of news organizations and other firms. Computer creativity will be explored more in later chapters in the areas of art, music, games, and more.

Decision Making

Artificial intelligence continues to impact decision making in every sector of society from the financial markets with algorithms trading stocks to the courtroom where judges refer to computers to suggest the appropriate sentences for defendants. Sally Percy in an article titled, "Artificial Intelligence: the Role of Evolution in Decision-Making," cites a decision-making process developed by the United States Air Force fighter pilot and strategist John Boyd (1927-

---

[18] Emily Reynolds, "Wordsmith's 'robot journalist' has been unleashed," *Wired*, October 20, 2015, accessed online October 9, 2017

1997) called *OODA*.[19]   OODA stands for Observe, Orient, Decide, and Act and is the decision loop he developed that revolutionized the success rate for US fighter pilots. He found that the pilots who applied this principle most frequently were the most successful. Moreover, OODA applies to any process where survival depends on making decisions in dynamic changing systems. The OODA loop has since been applied across the military, beyond to corporations, and in artificial intelligence. "The theory of the OODA loop is central to the activities of Sentient, which is the world's most funded artificial intelligence (AI) company."[20] Large data, ultra-fast computing, and rapidly changing environments lend themselves to artificial intelligence applications using the OODA loop in many sectors from financial markets to personalized customer experiences.

The future envisioned by companies such as Apple, Google, General Motors, and Daimler-Benz, shows people whisking around in driverless cars that use advanced sensing capabilities based on artificial intelligence, robotics, and advanced decision-making algorithms. The driverless cars will not be subject to human error, lapses in judgment, slow reflexes, or failing eyesight. For a driverless car to succeed, it needs to be able to determine its surroundings very accurately. Just as when you are driving a car, and you see something on the road ahead, you need to decide is that a person, animal, or just a plastic bag? Under normal driving conditions, you may slow down or change lanes if it is a

---

[19]   Sally Percy, "Artificial intelligence: the role of evolution in decision-making," *The Telegraph*, March 23, 2017, accessed online October 9, 2017
[20] ibid

plastic bag. But what if it is a person? For a driverless car to be successful, it will need to make these same decisions. Sometimes there will be decisions to make that no matter what, will have a negative impact. Take, for example, a person jumps into the road in front of a crowded bus stop. The driverless car cannot stop in time to avoid the person, and it will need to choose between veering left into oncoming traffic, causing a head-on collision, veering right into the crowded bus stop, perhaps hitting many people, or hitting the person who jumped into the road.

The belief that management positions, with their need for emotional as well as intellectual skills, will remain safe from automation should be challenged. Corporate and small business management must accept that their jobs as managers will change significantly as new tools powered by artificial intelligence work their way into offices and even C-suites. How can management delegate work to artificial intelligence where it makes sense, and what will management need to know to succeed with new AI tools and robotics? In many respects, AI and robotics will naturally displace humans from certain types of activities where analytical data drives decision making. Machines have the advantage of much faster analysis of data than a human. In "What to Expect from Artificial Intelligence," management experts envision a shifting role in management from making predictions to applying judgment.[21] As artificial intelligence improves in making predictions and as it also becomes more widespread and inexpensive, more

---

[21] Ajay Agrawal, Joshua S. Gaines, and Avi Goldfarb, "What to Expect From Artificial Intelligence," *MIT Sloan Management Review,* Spring 2017 Issue, vol. 58 No. 3, Reprint #58311

computers may assume more prediction roles. People perform predictions all day long, from estimating how long it will take to drive to a business meeting, which candidate will be best for the job, how best to allocate profits back into the business, and such. Under certain circumstances, some predictions can be handled by a computer. For example, one of the cornerstones of the banking system is the decision to offer credit or a loan. Embedded in that decision is the prediction of whether the applicant will at some point in the future pay back that loan. Artificial intelligence, using a wide variety of information, offers strong predictive capabilities in this area.[22] However, predictions alone are not the same as decisions, and judgment is necessary to make the right choice. "Judgement is the ability to make considered decisions—to understand the different impact actions will have on outcomes."[23] The loan officer's judgment will look at more than the predictions, but at the whole person and consider the value of accepting the risk of the loan. The challenge of the future will be developing people with good sound judgment to apply best the new predictive capabilities of artificial intelligence.

Interestingly, decision making when delegated to machines in some cases reduces bias and improves efficiency. In an article in the *Harvard Business Review* titled, "4 Models for Using AI to Make Decisions" by Michael Schrage, the author suggests that machines

---

[22] Shorouq Fathi Eletter, Saad Ghaleb Yaseen and Ghaleb Awad Elrefae, "Neuro-Based Artificial Intelligence Model for Loan Decisions," *American Journal of Economics and Business Administration*,2 (1): 27-34, 2010

[23] Ajay Agrawal, Joshua S. Gaines, and Avi Goldfarb, "What to Expect From Artificial Intelligence," *MIT Sloan Management Review,* Spring 2017 Issue, vol. 58 No. 3, Reprint #58311

will find a home soon in the executive suite of the companies that want to compete successfully in a more data-driven marketplace. Interestingly, he notes that "Executives who wouldn't hesitate to automate a factory now flinch at the prospect of deep-learning algorithms dictating their sales strategies and capex [capital expenditure]. The implications of success scare them more than the risk of failure."[24] It may be the case that machines will have access to more data than the CEO and make a better decision about capital expenditures than the boss. But does this mean that computers will be replacing the leadership? Probably not, but leadership will require different types of individuals that are capable of working with machines and better at applying judgment to the results.

Management may be benefiting from many new AI driven tools to assist in everything from employee recruitment, surveillance, and retention. The promise of new AI driven tools designed to relieve pain points for management and workers sounds promising, but managers and future managers will need to prepare for the new world. Looking forward, one must contemplate what type of skills management will need to adapt and succeed with AI, machine learning, and robotics. Management will need better skills in data analysis, an understanding of how AI works, including its limitations, and the ability to discern where human creativity and decision making works best for the organization. In an article in the *Harvard Business Review* titled "How Artificial Intelligence

---

[24] Michael Schrage, "4 Models for Using AI to Make Decisions," *Harvard Business Review*, January 27, 2017, accessed online October 22, 2017

Will Redefine Management," the authors identify key skills for new managers. "Managers we surveyed have a sense of a shift in this direction and identify the judgment-oriented skills of creative thinking and experimentation, data analysis and interpretation, and strategy development as three of the four top new skills that will be required to succeed in the future."[25]

Robots

Robots embody artificial intelligence, giving a computer the ability to act in the physical world. Through robots, computers not only make decisions, but they can then carry out actions. The translation of AI into the physical world through robotics offers both a massive potential for benefit and profit for humanity, but as artists and thinkers have warned, there is also a potential for harm to our jobs, safety, and even our way of life. Robotics covers a wide array of applications from large industrial robots able to move and manipulate massive objects, to human scale robots designed to help people with everything from medical and elder care down to nanobots that can be as small as a millionth of a millionth of a millimeter, or a nanometer. In popular culture, artificial intelligence (AI) emanating from computers and embodied in robots so often gets cast as evil or anti-human. Much less frequently, AI will be depicted as helpful and caring for humans. Notable examples include R2D2 and C3PO in *Star Wars* or Commander Data in *Star Trek: The Next Generation*.

---

[25] Vegard Kolbjørnsrud, Richard Amico, Robert J. Thomas, "How Artificial Intelligence Will Redefine Management," *Harvard Business Review*, November 9, 2016

Applications of Artificial Intelligence

Artificial intelligence is often described in movies, fiction, and even the news as a malevolent presence that will gobble up our jobs, destroy our privacy, and disrupt life as we know it. This chapter will cover some promising and some scary applications of artificial intelligence in robotics.

Karel Čapek introduced the term *robot* in his play R.U.R. (Rossum's Universal Robots) in 1920.[26] Robot comes from the Czech term *robota* that translates in English to *forced labor*. The prologue to the play sets the stage at a factory, and among other things such as a desk and phone, are posters on the wall advertising, "Looking to Cut Production Costs? Order Rossum's Robots." *Robot* means a programmable machine capable of carrying out complex tasks in a repeatable manner. But robot means so much more in our culture today not only from a manufacturing and automation perspective but in entertainment and even myth. As we consider the expanding relationship between computer science, robotics, and society, we must evaluate the possibility and problems of blending robotics and artificial intelligence

Providing humans with support in harnessing the ever-expanding knowledge space will amplify man's ability to leverage knowledge faster and better than ever before. The parallel combination of AI with robotics brings the accomplishments of computing out of the abstract world of the computer into the real world where AI helps robots move, interact with people and objects, make decisions, and perform actions on their own without human assistance. Robots are not

---

[26] Karel Čapek, *Rossumovi Univerzální Roboti* (Rossum's Universal Robots), January 25, 1923

the future; they are now. Robots of many types and purposes already fill roles in industry, health care, transportation, companionship, the military, and more. The news regularly updates the public on autonomous, self-driving cars under development by Google, Tesla, Uber, and others. Other autonomous vehicles are moving beyond the research and development stages in different segments of industry.

Autonomous Vehicles

Autonomous vehicles would not be possible without artificial intelligence to allow them to see, plan, and make decisions. The US Navy teamed up with the Defense Advanced Research Projects Agency, DARPA, after six years of collaboration, built and in April of 2016 christened the *Sea Hunter*.[27] The *Sea Hunter* is the first large-scale autonomous submarine hunting ship dubbed *ACTUV*, or the Anti-Submarine Warfare Continuous Trail Unmanned Vessel. Using advanced sensors and artificial intelligence, *Sea Hunter*, a diesel-powered ship, was designed to navigate the waters of the world on its own without the close support of other ships or people. *Sea Hunter* is a 132-foot-long diesel-powered trimaran that holds enough fuel for it to spend months at sea hunting submarines without human intervention. *Sea Hunter* is not a remote-controlled drone run by sailors remotely. *Sea Hunter* uses advanced sensors and artificial intelligence to navigate the ocean. *Sea Hunter* has proven in trials that it can obey the

---

[27] Scott Littlefield, "Anti-Submarine Warfare (ASW) Continuous Trail Unmanned Vessel (ACTUV)," Defense Advanced Research Projects Agency Program Information, Accessed online January, 2017

nautical rules for navigation and anti-collision in all conditions of weather twenty-four hours a day. *Sea Hunter* must comply with International Regulations for Preventing Collisions at Sea (COLREGS). For an autonomous vessel, much like an autonomous car, *Sea Hunter* will use radar and other image processing software to determine not only the presence of other ships but also what the type and identity of other objects on the sea. Also, to find and track submarines, even the very quiet diesel-electric types, *Sea Hunter* has sophisticated sonar and other detectors that allows it to tail an enemy submarine for weeks. Sea Hunter can relay information back to command centers and even work with other detection systems like sonar buoys. Working in concert with other Navy assets, *Sea Hunter* and other ACTUVs extend the reach of the Navy while significantly lowering the cost of tailing submarines.

An article published by the Department of Defense quoted Scott Littlefield of DARPA's Tactical Technology Office that *Sea Hunter* would cost around $15,000-$20,000 per day to operate compared to $700,000 per day for a fully manned destroyer.[28] The substantial cost savings of the *Sea Hunter* strongly argues for the increased deployment of the ACTUV technology. Although named the *Sea Hunter*, it is a tracker and does not carry any weapons that would make it a true hunter.

Currently, the US Navy does not permit weapons on autonomous ships, but the nature of sub hunting is dangerous. It seems tempting to provide some self-defense capabilities to the Sea Hunter and other

---

[28] http://science.dodlive.mil/2015/11/09/actuv-sea-trials-set-for-early-2016/, accessed November 19, 2016

autonomous vessels. After all, roaming the sea looking for submarines makes such ships filled with advanced technology natural targets for enemies and pirates. It follows that if autonomous ships suffer damage or destruction from hostiles that the navy will eventually want to equip the unmanned ship with the ability to defend itself in dangerous situations. In a similar vein, autonomous cars that have been under development for years also face the need for self-protection.

Autonomous ships plying the oceans through the dark of night and relentless storms following hostile submarines sounds like a distant idea far from ordinary life, but the far more immediate autonomous vehicles that already impact our lives are the autonomous/driverless cars being developed today by corporations such as Google, Apple, and Tesla. Current reports indicate significant progress towards the first truly driverless cars.

A driverless car combines a dazzling array of sensors, communication, and computing systems for the vehicle to travel through a complex world that it does not know or value the same way a person does. According to Google, its autonomous car has logged over 2 million miles of accident-free (except for a no-fault time the vehicle was rear-ended) driving on public roads.[29] The car uses Laser Illuminating Detection and Ranging, also called LIDAR, to build a 3D picture of the world around it. By bouncing laser light off of objects all around the car, the computer can *see* other vehicles, curbs, objects in the road and even people up to two football fields away. But what LIDAR does not

---

[29] *https://www.google.com/selfdrivingcar/ accessed on November 20, 2016*

do well is figuring out how fast everything is moving. To better know how fast it is going and all the things around it like other cars, bicyclists, and pedestrians, the car has four radar units mounted on its four corners that talk to the onboard computer. Add to this bristling array of sensors a camera that reads traffic signals, street signs, and figures out the difference between a bird in the road or just a crumpled piece of newspaper. The Google car computer has navigated streets in California, Washington, and Texas.

The Google car is not available yet to consumers, because it has some significant technological hurdles to overcome such as a lack of detailed 3D maps the car uses to navigate roads. The Google car does not use the Google maps most people use with their smartphones but a highly detailed map of everything along a stretch of road down to the height of all the curbs and the exact width of the lanes. In fact, the Google test roads cover less than 0.1% of the public roads in the United States. There is much more mapping to be done and further technical hurdles such as being able to accurately tell the color of a stop light under high glare situations before the Google car replaces all of our regular cars.

Unlike the more cautious Google, Tesla has audaciously outfitted and sold a number of their electric vehicles with autopilot capacities that require the driver to set the desired speed and let the car do the driving. In essence, Tesla, using sophisticated sensors and data processing, put a very advanced cruise control standard in all their cars since 2014 that they call *autopilot*.[30] Tesla first activated the autopilot capabilities with

---

[30] https://www.tesla.com/presskit/autopilot#autopilot Accessed November 24, 2016

over-the-air software upgrades in October 2015. The Tesla, like the Google car, uses a number of sensors and cameras combined with AI in the onboard computer to navigate the vehicle completely hands-free. Although Tesla claims full autonomy will make their cars safer than human drivers, there are still issues to be solved. In May of 2016, an Ohio man died when his Tesla on autopilot slammed at 75 miles per hour into a tractor-trailer loaded with blueberries on a Florida highway. The top of the car was completely torn off. Tragically, the autopilot failed due to glare coming off the truck which confused the sensors. Tesla in a statement pointed out the very high safety record for their autopilot in over 200 million miles of driving. The statement sounds defensive in light of the loss of life. Even though fatalities occur with human-operated vehicles, the sense of risk and danger when handing control over machines challenges us as people in our sense of control, usefulness, and capability.

Robot Surgeon

Artificial intelligence is not only being entrusted to safely drive people and products around but with something more intimate—our hearts. In 2006, Dr. Carlo Pappone MD, Ph.D. performed the first transatlantic heart surgery, an arrhythmia ablation procedure, from Boston successfully, while the patient was 4,000 miles away in Milan, Italy.[31] The robot surgery, controlled by Dr. Pappone using a computer keyboard,

---

[31] Carlo Pappone, MD, PhD, FACC, Biographical Sketch, http://www. af-ablation.org/uploads/cv-carlo-pappone.pdf , accessed online February 16, 2017

demonstrated the power of robotics and the possibility of a remote operation. He used Intuitive Surgical's da Vinci surgical robot that allows steadier, more precise movements than the human hand and works through tiny incisions, removing the need for more invasive open-heart surgery. Robots assist in various other types of surgery from knee and hip replacement with the Mako robotic arm[32] to brain and spine surgery with robotics from Zimmer Biomet. The next stage in robotic surgery will use artificial intelligence to do fully autonomous surgery. It was announced recently that "In a robotic surgery breakthrough, a bot stitched up a pig's small intestines using its own vision, tools, and intelligence to carry out the procedure. What's more, the Smart Tissue Autonomous Robot (STAR) did a better job on the operation than human surgeons who were given the same task."[33] Hard tissue (tissue having a firm intercellular substance, like cartillage and bone[34]) surgery lends itself better to robotics because a rigid subject maps easily. However, soft tissue (body tissue except bone, teeth, nails, hair, and cartillage[35]) surgery poses a more significant challenge to map because the tissue moves and is often all the same color, which is why STAR's achievement with the soft tissue of an intestine marks an extraordinary breakthrough for artificial intelligence and robotics.

---

[32] https://patients.stryker.com/hip-replacement/procedures/mako-robotic-arm-assisted
[33] Eliza Strickland, "Autonomous Robot Surgeon Bests Humans in World First," *IEEE Spectrum*, May 4, 2016, accessed online October 10, 2017
[34] hard tissue." *Farlex Partner Medical Dictionary*. 2012. Farlex 4 Apr. 2018 https://medical-dictionary.thefreedictionary.com/hard+tissue
[35] "soft tissue." *McGraw-Hill Concise Dictionary of Modern Medicine*. 2002. The McGraw-Hill Companies, Inc. 4 Apr. 2018 https://medical-dictionary.thefreedictionary.com/soft+tissue

Elements of heart surgery that require precision and that need to accommodate variation between hearts in different patients pose a considerable challenge for the robotic surgeon. However, just as AlphaGO learned to play GO by playing itself, the next generation of robotic surgeons can learn from watching thousands of surgeries and practicing virtual operations before touching a human.

Swarm Robotics

Many robots have been designed to look and act like people because we often are looking for ways to automate what people do. However, the field of robotics and artificial intelligence also gets inspiration from other animals in the natural world. Although many people find insects creepy and even frightening when thinking of them in swarms such as fire ants or killer bees, entomologists study insects and continue to seek to understand how ants and bees function as a collective group with no central organizing leader. In most cases of social insects, the colony serves the queen, but the queen does not have a group of dedicated administrators that tell each ant or bee what to do. In the case of ants, individuals will respond to their environment through chemical signals called pheromones emitted by other ants they encounter. For example, an ant that has returned from a food source will emit a pheromone combination that tells another ant to start gathering food. The second ant without food will follow a trail of scent laid down by the first ant back to the food. Ants are versatile, and each ant can perform tasks like foraging, shuttling baby ants around, nest building, and more. Following simple signs from one individual to another gives the colony a unique ability to thrive without instructions or commands coming from some command center. The term *swarm intelligence* describes the ability of ants to build massive colonies, avoid danger, and invade whole new regions without a central leader.

Swarm intelligence refers to the amplified intelligence of groups of animals. Birds form flocks, fish form shoals, and ants form colonies. An individual animal like a fish or an ant may not be brilliant individually, but an animal swarm swarm avoids predators and builds magnificent homes like the termite colony or honey bee hive. The nature of swarm intelligence not only intrigues biologists but has captured the imagination of computer scientists and robotics engineers as well. Inspired by the collective intelligence of a colony of ants or bees, robotics research continues to delve into learning how simple robots can be programmed to work like a colony. They want to make robots that react to their environment and each other as a collective. This research forms a new branch of AI called *swarm intelligence*. Swarm sounds frightening and menacing, but some applications being developed today that use swarm intelligence may prove very helpful in areas such as agriculture and healthcare. In the agriculture industry, David Dorhout believes that the future of agriculture may depend on armies of cheap little robots equipped with sensors and some basic rules to replace massive farm machinery. He envisions little robots using swarm intelligence to plant, tend, and harvest various crops. There will not be some complicated central planning programs that tell the robots where to go and plant. Instead, the robots will sense where other bots are and move to open spaces to plant seeds and detect when moisture is low and water individual plants. Mr. Dorhout calls his company Dorhout R&D and his farm robot Prospero. The medical community also envisions employing swarm intelligence at the level of nanoparticles for treating cancer. The tiny nanoparticles will pass through a blood vessel into cancerous tissue, recognize a tumor, and destroy it. Swarm intelligence may also contribute to making autonomous cars safer by letting cars talk to each other locally to prevent collisions and smooth traffic flow.

Clearly on the more menacing side, military applications of swarm robotics may come in many forms. The military frequently leads the way in technological advancement and already supports numerous projects developing robotics and swarm technology. According to an article in *New Scientist*, the US military envisions using robot swarms in removing mines and for search and rescue without human intervention.[36] The US Navy released a video on YouTube of F/A 18 Hornet fighter jets launching just over 100 mini-drones over the desert and the subsequent testing of the drone swarm.[37] Unlike typical drones that have one operator per drone, the drone swarm uses a collective brain shared among all the drones that communicate and works like a swarm in nature. The drones are battery powered, 12-inch-long aircraft named Perdix and were developed by students at Massachusetts Institute of Technology's Lincoln Labs. They are made of commercially available

---

[36] Will Knight, "Military robots to get swarm intelligence," *New Scientist*, April 25, 2003, accessed online August 6, 2016
[37] "US Fighter Jets Launch Drone Swarm of Hundreds of Micro Drones: Perdix Micro-UAV Drone Swarm Test," US Navy, YouTube, https://www.youtube.com/watch?v=5NGgHyfPGU0. Accessed July 2017

and 3-D printable parts. The video also shows the swarm clustering and then moving to patrol an area. They look and sound like a menacing swarm of huge bees or hornets. It seems almost ironic that the Hornet was used to launch in flight the drones from underwing pods. The Perdix drones may be initially helpful for surveillance in battlefield situations. By extension, their small size, the large number, and low cost could also make them a new and challenging to defend against weapons.

Swarm intelligence observed in nature in the form of ant and bee colonies demonstrate how the actions of individuals together without a central organizer amplify their collective intelligence that is smarter than any individual ant or bee. The collective intelligence observed in the natural world inspired a new discipline in artificial intelligence that seeks to use cooperative behavior to empower simpler robots to work as a team for a common goal.

Conclusion

Artificial Intelligence covers many areas in which computers display traditionally human capabilities such as perception, language processing, planning, creativity, and decision making. To perceive is to see, to touch, to smell, in short, the ability to know what and who is out there, and artificial intelligence gives this ability to machines. Furthermore, computers are now learning more about what they are perceiving, which makes them more useful for planning and ultimately for decision making even in the most crucial situations. Moreover, robotics provides the embodiment of artificial intelligence. Through a robot, artificial intelligence can interact with the physical world like a person. Without artificial intelligence, breakthrough technologies from autonomous vehicles, to robotic surgeons and agricultural swarm robots would not be possible. Having machines that can learn opens up a vast array of possibilities for innovation. We will explore the relationship between AI and robotics as we look at how to profit and protect ourselves from these emerging technologies.

Chapter 4
State of Artificial Intelligence Today

"Just what do you think you are doing, Dave?"

—Hal, the computer from *2001: A Space Odyssey*

Artificial intelligence already surrounds us. We are surrounded by computers almost everywhere we go these days, and many of them already use artificial intelligence. Obviously, the smartphones, laptops, and tablets everyone uses are computers that have artificial intelligence, but there are computers in our cars that manage many functions like optimizing engine function and operating your anti-lock brakes, the backup camera, and the bumper sensors. New smart televisions contain a computer to run the digital screen and the multiple inputs, including the internet and other video streaming content. Computers also appear in many kid's toys. For example, the once wildly popular Furby produced by Tiger Electronics and originally launched in 1998 was an animatronic robot capable of voice recognition with embedded computer programming that allowed it to interact with people and other Furbys via an infrared sensor. New toys today are being released that use artificial intelligence to customize your child's playtime. Cognitoys claims that their smart toy called Dino, with the help of IBM's Watson supercomputer, will use artificial intelligence to become your child's friend, teacher, and companion.[1] Through Wi-Fi, Dino can answer thousands of age-appropriate questions, make jokes, and, through artificial intelligence, adapt to each child's needs and abilities. The presence of computing will undoubtedly grow, especially with the burgeoning *internet of things*. The architects of the internet of things claim the internet of things will produce and receive information from every conceivable place from machines to plants and animals to structures to even you. The internet of things is already growing in the use of internet-controlled lighting to the gathering of human vital signs being conveyed to computers already through smartwatches, and even internet connected pacemakers.

The powerful, ever present computer often known just from merely its voice has played a role in many science fiction movies, TV shows, and novels. Many may recall Captain Kirk calling the ship's computer in *Star Trek* to provide helpful data or analysis to save the day. It seemed like a remarkable fantasy. Depending on the genre, the computer may be helpful like in *Star Trek*, funny like C3PO in *Star Wars* or, in the case of Hal in *2001: A Space Odyssey*, sinister and above man.

---

[1] https://cognitoys.com/

The computer may be silly or even protective, but the concept of an omnipresent source of intelligence has fascinated man for much longer than the idea of the computer. Yet in a less dramatic form, we live with computers among us today that we talk to and interact with all the time. From robotic call screeners on the phone to digital assistants on our smartphones or in computers and systems that find us the quickest route to get to a restaurant across town, we are already living with computers and robots today. In this chapter, we will look into notable examples of computers and robots that impact our lives now and not just in science fiction. Furthermore, we will look at artificial intelligence as we encounter it through digital assistants such as Siri, supercomputers like Watson, emotionally intelligent robots, socially intelligent programs, and our aspirations and fears about artificial general intelligence.

Siri

Anyone who has a computer or smartphone has access to or been pestered by the voice-activated digital assistant called Siri in Apple products or Cortana in Microsoft's Windows or Android. Siri is designed to be the ubiquitous digital assistant that brings knowledge and action to you by merely talking to your device. You can ask what is the best steakhouse in town, and Siri will use artificial intelligence to come up with the name of the restaurant and even tell you how to get there. Other companies have competing digital assistants, but I will focus on Siri as an example to explain what this system is and how it is penetrating our lives.

Siri first appeared in October 2011 with the Apple iPhone 4S. Since then, Siri has been integrated into many Apple products from tablets to Apple TV. As a digital assistant, Siri combines voice recognition and natural language processing to understand and adapt to an individual's voice and responds with individualized, comprehensible verbal answers and commands. Siri not only needs to understand human voice commands but needs to process the information to formulate responses. All of this complexity masked by the simple voice interaction draws on other products and applications such as the Maps app, reservation systems like OpenTable, question answering services, and more.

Siri did not begin at Apple. Instead, Siri started back in 2003 as a research project funded by the Defense Advanced Research Projects Agency, or DARPA. DARPA, created by President Eisenhower in 1958, is an agency of the US Department of Defense charged with the mission to develop emerging technologies for military purposes. DARPA wanted to build an intelligent virtual assistant that would help military personnel be more organized. The goal of creating virtual assistants serves, according to the US Information Innovation Office, "To make information understanding and decision-making more effective and efficient for military users."[2] The project had a massive goal to use artificial intelligence to make a virtual assistant that learns as it goes, personalizes its responses, and responds to unpredictable situations. The project was called Cognitive Assistant that Learns and Organizes (CALO) under the Personalized Assistant that Learns (PAL) initiative. SRI International, the

---

[2] https://web.archive.org/web/20110805162949/http://www.darpa.mil/ Our_Work/I2O/Programs/Personalized_Assistant_that_Learns_(PAL). aspx accessed November 5, 2016

non-profit R&D organization that had spun off from Stanford in the 1970s, was given the role of coordinating the hundreds of computer scientists, engineers, and researchers that were responsible for CALO. It also later was the inspiration for the name *SIRI* from *SRI International*.

CALO set out to become a super assistant with the capability to access calendars, emails, manage appointments, integrate information from many sources, and make decisions. In the words of Adam Cheyer, the Chief Architect of the CALO project at SRI, "The goal was ambitious: to bring together all 'stove-pipe' aspects of artificial intelligence into an integrated, human-like system that could learn in the wild. With no code changes, the system would get smarter over time by observing the user, interacting with him or her, and self-reflecting on what it saw and heard."[3] CALO would learn the roles of the people in meetings or on emails or presentations. CALO would know if higher ranking officers were scheduled for a meeting and, if the meeting had to be canceled, suggest the best time to reschedule. CALO could also assemble documents and presentations based on past presentations, reports, and new relevant information. Like a highly skilled administrative assistant, CALO could record meetings, take minutes, note action items, and even provide preparation for upcoming meetings. Moreover, CALO could be taught routine tasks by interacting with its user and also manage schedule prioritizing based on experience. The CALO project set lofty goals and

---

[3] From an interview with Adam Cheyer by Danielle Newnham, https://medium.com/swlh/the-story-behind-siri-fbeb109938b0#.5b99vz7x9, accessed November 6, 2016

proved worthy of the job. After five years and a relatively modest $150-250 million investment, many of the goals were achieved, and CALO was deployed by the US Army in Iraq in 2010 as a component of its Command Post of the Future (CPOF).

The CALO project proved to be a massive success with breakthroughs in artificial intelligence and machine learning, and some companies were spun out from the project, including Siri Inc., Tempo AI, and Kuato Studios, which develops educational games. Siri Inc. initially provided Siri as an app for the iPhone available through the App Store, but soon after the app was available, Apple acquired Siri Inc. in the spring of 2010. In the following year, Siri was integrated into the iPhone and expanded to work with many foreign languages from Chinese to French to Arabic with more improvements to deal with dialects and accents. Interestingly, an article in the *Huffington Post*[4] implied that the Siri we get from Apple is only a fraction of CALO and that many of the capabilities of CALO were turned off for Apple's Siri. In fact, CALO had become able to learn on its own, without additional programming, the interests and preferences of its users, combine information from multiple sources, and execute a request. The engineers at CALO called it a *do engine* which was the next generation of the search engine like Google or Bing. The *do engine* would be the assistant that gets you and seems to know what you need just before you do. Even if Siri is only a shadow of what

---

[4] Bianca Bosker, "SIRI RISING: The Inside Story Of Siri's Origins — And Why She Could Overshadow The iPhone," *Huffington Post* http://www.huffingtonpost.com/2013/01/22/siri-do-engine-apple-iphone_n_2499165.html, accessed November 5, 2016

it was at first, that does not mean that improvements are not already happening and that the competition is fierce.

Other AI assistants already occupy space in the market such as Amazon's Echo and Microsoft's Cortana. Echo comes as a cylindrical, voice-activated smart speaker for home use that responds to the name *Alexa*. Echo contains speakers and microphones that connect to the internet to deploy its voice recognition capabilities to play songs and check traffic and the weather just by calling out Alexa's name. You wake Echo by calling out her name, *Alexa*, and the system next transmits what you say to Amazon's AI computers for analysis, and then Amazon relays its response back to your Echo. The commands can be more than requesting a song or the news; Echo can also interact with smart devices in the home like smart lights and even order products or a pizza if you set it up. Some of the commands for Echo are very specific, so you need to know how to talk to Echo. You need to know the command to skip the current song or how to get the news. Apart from some shortcomings, having a hands-free voice-activated computer in your house that performs tasks on command feels more and more like having that ubiquitous computer in *Star Trek*.

Watson

On February 14-15, 2011, the world watched IBM's supercomputer, Watson, win the question-answering game show *Jeopardy!* by soundly defeating two of the show's most accomplished champions, Brad Rutter and Ken Jennings. In the defeat, Watson earned three times more money than the nearest human competitor. If the loss of the world champion chess player, Gary Kasparov, in 1997 to IBM's supercomputer Deep Blue hailed the first triumph of the computer over man, the *Jeopardy!* win by Watson marked the second significant milestone. People watched in amazement as the computer would listen to the questions being read and answer them in its not exactly human voice.

Watson was built as a question-answering computer specifically to compete on *Jeopardy!*. Behind the voice and logo visible to the public on the game show, IBM computer scientists had made a very sophisticated and powerful computer. Watson represents an essential step in the development of artificial intelligence because of the way that it works. Watson is not a single computer program that runs on a machine. Instead, Watson uses many different computer programs to achieve the correct answers it produces. Watson has programs to analyze the written word in books, dictionaries, newspapers, on the web such as Wikipedia, and other electronic records. It also has applications to recognize and understand the spoken word as well as the ability to reply in English for people to understand. Moreover, Watson uses many different programs to generate possible answers to the questions it gets asked. Instead of searching for a solution in a set of known responses, Watson generates many possible answers by breaking the question down into parts and looking for connections in the millions of pages of information it has in its memory. Using different programs running at the same time, Watson generates many possible answers. Next, it identifies the best answers and then ranks these responses against similar answers it knows are correct. Herein lies the power and the limitation of Watson. The part of Watson that has known right answers was trained by humans. People need to teach Watson on what is correct or incorrect. With a good set of known, correct answers, Watson can make excellent guesses. Over time, as Watson develops experience, it can learn from its mistakes as well.

Since the *Jeopardy!* win, Watson has been used for many question-answer type applications. In California, water is a precious commodity, especially in agriculture. Watson and the E&J Gallo Winery teamed up to provide precise vine by vine irrigation to optimize wine quality and water conservation.[5] Watson analyzes weather patterns, satellite images, and soil moisture to water just the vines that need it. The result has been better grapes and a 25% reduction in water use. Watson also helps oncologists make treatment decisions for their cancer patients. At the Memorial Sloan Kettering Cancer Center in New York, NY, teams of physicians have worked to train Watson on the best cancer treatments for a patient based on some symptoms and clinical information. Watson also consumes thousands of new cancer research papers and clinical studies from around the world.[6] The training from cancer specialists is helping Watson Oncology to become a valuable resource in treatment recommendations. Because Watson requires training, not everything works so well. In a publicized move, the MD Anderson Cancer Center in Texas ended their collaboration with IBM Watson after several years and millions of dollars spent on an

---

[5] "Can satellites help E. & J. Gallo Winery save water and produce better-tasting wine?," IBM Watson, https://www.ibm.com/watson/stories/ejgallo-with-watson.html, accessed online September, 2017
[6] Watson Oncology, Memorial Sloan Kettering Cancer Center, https://www.mskcc.org/about/innovative-collaborations/watson-oncology, accessed online September 2017

attempt to get Watson to help cure cancer.[7] In some cases, there are not enough right answers for Watson to work with and it cannot, even with its massive computing power, make useful recommendations to the scientists trying to cure cancer.

IBM's question-answering supercomputer Watson wowed the world when it triumphed in *Jeopardy!* on television. A system of sophisticated computer programs allows Watson to make educated guesses to problems in any topic that it has been trained in such as science or general trivia; however, Watson needs a lot of care and teaching from experts to reach its full potential. It is not all-knowing and needs to be pointed in the best directions to help people solve our most significant problems.

Emotionally Intelligent Machines

Alan Turing, back in the 1950s, suggested that for a machine to be truly intelligent, it would have to be able to hold a conversation with a person, and the person would not be able to tell she was having a conversation with a machine. For such a communication to work, it would require the machine to speak in a conversational tone and be able to pick up unspoken references, intonations, irony, and jokes. The computer would need to understand human emotion. In a now considered foundational paper, "Affective Computing," written in 1995 by MIT professor, Rosalina Picard, issues related

---

[7] Matthew Herper, "MD Anderson Benches IBM Watson In Setback For Artificial Intelligence In Medicine," *Forbes*, February 19, 2017, accessed online September 2017

to emotion in computing are outlined and discussed.[8] She discussed growing the capability in computers to detect emotion, react to emotion, and even include emotion in computer decisions. The paper effectively combining computer science, neurology, psychology, and philosophy launched the field of *Affective Computer Science*. *Affective Computing* challenges how we can expect to relate to and interact with computers now and increasingly in the future. Some of it gets genuinely philosophical. From the perspective of artificial intelligence, Dr. Picard not only suggests the likelihood of computers acting in ways to pull emotions from us but also sensing our emotional states to tailor how the computer should interact with us.

Dr. Picard founded a company called Affectiva with Rana el Kaliouby of the MIT Media Lab in Cambridge, MA. Affectiva develops emotionally intelligent AI intended to help computers respond to verbal and non-verbal emotional signals. Dr. Kaliouby heads the emotion analytics team developing these emotion sensing capabilities. In a TED talk on June 15, 2015, she described how her and her team used 2.9 million face videos from 75 countries around the world to train the computer to read emotional states.[9] Volunteers watching funny videos labeled their emotions, which made the training set for the computer.[10] The volunteers

---

[8] Rosalind W. Picard, "Affective Computing," M.I.T Media Laboratory Perceptual Computing Section Technical Report No. 321, 1995
[9] Rana el Kaliouby, "This App Knows How You Feel—By the Look on Your Face," TEDWomen 2015, May 2015, Accessed online April, 2017
[10] Daniel Jonathan McDuff, Rana el Kaliouby, Thibaud Senechal, May Amr, Jeffrey F. Cohn, Rosalind W, Picard, "Affectiva-MIT Facial Expression Dataset (AM-FED): Naturalistic and Spontaneous Facial Expressions Collected In-the-Wild," DSpace@MIT, http://hdl.handle.net/1721.1/80733, accessed

looked through pictures identifying the emotions they saw and then tagged the photos. The tens of thousands of tagged pictures combined with the facial recognition software and deep learning trained the computer to identify emotions from a picture, emotions such as happiness, sadness, worry, fear, and joy.

The other, more controversial, side of the coin involves giving machines emotions. Picard notes that emotion from a neurological perspective plays a crucial role in human cognition. Emotions are not seen as separate or in opposition to learning and comprehension but essential to the process. Computer scientists may want to try to give emotions to machines so they can relate better to people, but what will it mean to how people relate to computers if the computer feels angry with someone? What if a computer feels threatened? How should it react?

Socially interactive robots continue to enter into society at an ever-increasing pace. One of the leaders in socially interactive robots, SoftBank (SFTBF), offers an artificial intelligence-enabled robot named Pepper. Pepper is a humanoid, or human-shaped, robot that stands 4-feet-tall with a smooth plastic head and large expressive eyes that change color with different emotional states. Pepper does not have two legs but rather a stem that widens at the base like feet with hidden wheels. Pepper rolls across level surfaces at a top speed of 2 miles per hour, which is slightly slower than the average human's walking speed. According to SoftBank.com, Pepper is the first commercially available robot that can react to people's emotions. Pepper learns the emotional state of people through their vocal tone, speech, and facial expressions.[11] Additionally, through advanced robotics, Pepper not only displays appropriate emotional responses through speech and changing eye color but through very human-like body language. Pepper's likeability and sociability make it a natural choice for commercial interaction situations as well as for personal companionship.

First commercially available in Japan in 2015, Pepper already occupies some roles in Japan's society. Pepper works at hundreds of SoftBank mobile phone outlets greeting customers and answering questions. Nestle, the multinational food and beverage corporation, employs Pepper in a thousand stores across Japan to help sell their coffee makers and coffee. SoftBank showcases Pepper on their website in a portrait of a multi-generational Japanese family as the newly adopted member of their family. More recently, as reported by the *Japan Times*, Pepper now works as a robotic Buddhist monk assisting in Buddhist funeral ceremonies. Dressed in traditional robes, Pepper interacts with mourners and chants sutras (Buddhist holy texts) from four different Japanese Buddhist sects. The company, Nissei Eco of Japan, programed Pepper to provide an inexpensive alternative to the traditional Buddhist funeral service with a monk in attendance who will chant sutras as part of the ceremony. Nissei Eco claims that Pepper costs significantly less at 50,000 Yen (about $457.00) compared to a monk at ten or more times the price. With funerals costing millions of yen (thousands of dollars), Nissei Eco hopes to fill

---

[11] Who is Pepper?, SoftBank.com, https://www.ald.softbankrobotics.com/en/robots/pepper, accessed on online August 25, 2017

a need for a dignified funeral at a reasonable price. In Buddhism, death occupies a significant part of the cycle of reincarnation. Although the funeral traditions vary from sect to sect, according to "A Guide to a Proper Buddhist Funeral," at the funeral, a monk chants sutras to release the good energies of the deceased from the body. The sutras help transit the person to their new life with the hopes of a good reincarnation. The reciting of the sutras helps the deceased into their next life. However, due to location or financial circumstances, monks may not be present for a Buddhist funeral, and it is acceptable for family members or other mourners to recite sutras instead. With modesty as an overriding principle in a Buddhist funeral, pre-recorded sutras fill an optional role in the funeral ceremony.

Another interesting example of the combination of robotics, artificial intelligence, and Buddhism comes from China. Recently, in a monastery outside of Beijing, *The Guardian* reported Xian'er, a 2-foot tall robotic monk developed by the Longquan Temple in collaboration with several Chinese tech companies to answer questions relating to Buddhism from the visiting public.[12] The Longquan Temple has an animation group and an information technology team with the goal of reaching the growing number of grassroots Buddhists that may cause the proliferation of misunderstandings of Buddhism. The expressed intent is to reach more people interested in Buddhism by embracing the digital age.

Emotionally intelligent robots continue to fill

---

[12] Harriet Sherwood, "Robot monk to spread Buddhist wisdom to the digital generation," *The Guardian*, April 26, 2016, accessed online April 17, 2017

more complex roles in society from concierge to personal companionship. Moreover, emotionally and artificially intelligent robots are finding their way into Buddhism in Japan and China. To another degree, we are seeing the improvement of technology and the simultaneous inclusion of emotionally intelligent robots into everyday life even to the most personal realm of religion. Perhaps, one day just as Netflix or Amazon get better and better at guiding our preferences and purchases, we may rely on AI in our more personal, spiritual guidance.

Artificial Intelligence, Problem Solving, and Bias

People regularly talk about the future and try to predict what might happen. We worry, we fret, and we try to get an advantage. Millions upon millions of dollars fund efforts to better predict stock market performance or housing prices. Just think of the amount of print and airwaves that are spent on trying to predict who the next president will be. During an election year, we listen to pundits and prognosticators tell us who will win the election, and they may use statistics from poll numbers or their experience or to relate the sentiment of the people. The policy advisor and political consultant, Karl Rove, who served as the Senior Advisor to the President and Deputy Chief of Staff for President George W. Bush from 2001-2007, is credited with the election victories of many political candidates, including the gubernatorial and presidential elections of President Bush. After resigning from his White House positions, Rove moved on to political analysis working for several publications and news organizations. Famously, Karl Rove predicted that Mitt Romney would win the general election for President of the United States, but the respected analyst was wrong. On Halloween 2012, he wrote in the *Wall Street Journal*, "It comes down to numbers. And in the final days of this presidential race, from polling data to early voting, they favor Mitt Romney."[13] Rove cited past election trends and poll numbers and made comparisons to the past, but in the end, Barack Obama defeated Mitt Romney in both the popular and the electoral vote to be President for a second term. The future is uncertain, but as Daniel Kahneman, the Nobel Prize-winning economist, has noted, *people are not good at intuitive statistics based on our own biases*.[14] Even Karl Rove, a trained statistician, ignored the polls showing the growing momentum for president Obama. Various types of AI and machine learning tools apply statistical models to massive amounts of data to predict the outcome of many events, including presidential elections.

Artificial Intelligence models take vast amounts of data and render predictions that have a less human bias. A great strength of experts can be their insights and perspective in helping to solve problems. However, expertise at times contains bias that obscures the real truth. Machine learning has emerged as a powerful tool to see beyond the bias with which even experts may struggle. In the presidential election of 2016, the pundits and the polls suggested a tight but convincing win for Secretary Hillary Clinton. Polls at the local

---

[13] Karl Rove, "Rove: Sifting the Numbers for a Winner," *Wall Street Journal*, October 31, 2012 http://www.wsj.com/articles/SB10001424052 970204846304578090820229096046 accessed November 6, 2016

[14] Daniel Kahneman, *Thinking, Fast and Slow*, Farrar, Straus and Giroux, New York, New York, 2011

level, at the state level, and the national level missed the election.[15] Respected polls and pundits from CNBC, *Wall Street Journal*, and Monmouth University missed the outcome. In contrast, machine learning predictions relying on data such as social media attention accurately predicted the result of the election sometimes even months out from the final vote. For example, the artificial intelligence tool MoglA, developed by Sanjiv Rai, has correctly predicted each presidential race in the US since 2004 and even predicted the primary outcomes in 2016. In the most recent presidential election, MoglA predicted Donald Trump would win weeks before and right up to the election against all the polls and opinions in the media. CNBC reported ten days before the election that artificial intelligence was seeing a very different outcome, and none of the experts believed it. In contrast, the respected statistician and writer Nate Silver gave Hillary Clinton a 73% chance of winning the election as reported on the news website *FiveThirtyEight*.[16] On election day, the *New York Times* gave Secretary Clinton an 85% chance of winning the election,[17] and the *Huffington Post's* poll of polls gave Clinton a 47.3% to 42.0% lead.[18] The polls somehow missed the mood, but ironically just

---

[15] Carl Bialik and Harry Enten, "The Polls Missed Trump. We Asked Pollsters Why." Fivethirtyeight.com, November 9, 2016, http://fivethirtyeight.com/features/the-polls-missed-trump-we-asked-pollsters-why/ accessed on November 12, 2016

[16] http://projects.fivethirtyeight.com/2016-election-forecast/; FiveThirtyEight accessed on November 13, 2016

[17] Jos Katz, Who Will Be President?" *New York Times,* 2016 http://www.nytimes.com/interactive/2016/upshot/presidential-polls-forecast.html; accessed in November 13, 2016

[18] 2016 General Election: Trump vs. Clinton, http://elections.huffingtonpost.com/pollster/2016-general-election-trump-vs-clinton; *Huffington Post* accessed on November 13, 2016

like Carl Rove in 2012, *The New York Times* assessment ignored the upward trending of Mr. Trump in many of the battleground states in preference for the overall numbers.[19] Was this another example of bias clouding judgment? MoglA works with the data in an unbiased manner using a model that looks at tens of millions of data points such as engagement on the internet with Facebook, Google, Twitter, and YouTube. The artificial intelligence system develops its own rules and continues to learn year after year what trends in social media mean for each candidate. MoglA has found that surging social media attention, positive or negative, correlates with election outcomes better than polls. Predicting election outcomes reflects only a fraction of the rapidly growing strength, quality, and impact of AI and machine learning systems on how we look at the world and anticipate change. Artificial intelligence is not always without bias, which will be looked at in later sections.

Artificial General Intelligence

The press, with the help of sensational movies and books about artificial intelligence such as *The Singularity Is Near*, depicts a technological revolution driving rapidly and inevitably to the point at which machines will eclipse man as the most intelligent entity on Earth. The singularity imagines a superintelligence that will be in a constant state of self-improvement, each improvement building on the last, creating an explosion

---

[19] Jos Katz, Who Will Be President?" *New York Times,* 2016 http://www.nytimes.com/interactive/2016/upshot/presidential-polls-forecast.html; accessed in November 13, 2016

of intelligence that will signal the end of humans as being the highest intelligence. Various theorists predict that the singularity will come. Stuart Armstrong of the Future of Humanity Institute at Oxford University in the United Kingdom in 2012, based on expert opinions, concluded that the singularity might arrive around 2040.

The growing concern about super intelligence emanates not only from some dramatic stories in the media but also key leaders and thinkers from around the world have spoken out too. Famously, the renowned astrophysicist, Professor Stephen Hawking, told the BBC back in 2014, "The development of full artificial intelligence could spell the end of the human race. It would take off on its own, and re-design itself at an ever-increasing rate. Humans, who are limited by slow biological evolution, couldn't compete, and would be superseded."[20] Hawking is not alone. Elon Musk, the technology entrepreneur, founder of the electric car company, Tesla Motors, and SpaceX, a private space exploration firm, has warned that "AI is our greatest existential threat." Following these comments, Musk, Hawking, and a large number of AI experts signed an open letter on artificial intelligence calling for research into the social impacts of AI. The letter written by the Future of Life Institute was titled "Research Priorities for Robust and Beneficial Artificial Intelligence: an Open Letter."[21] The letter acknowledges the potential

---

[20] Rory Cellan-Jones, "BBC News, Stephen Hawking warns artificial intelligence could end mankind," *BBC News*, December 2, 2012, accessed on line on March 11, 2017

[21] Stuart Russell, Daniel Dewey, and Max Tegmark, "Research Priorities for Robust and Beneficial Artificial Intelligence, Association for the Advancement of Artificial Intelligence," futureoflife.org, Winter 2015,

benefits of super-intelligent machines to help solve some of man's most significant problems such as disease, poverty, and hunger as well as noting the real danger of AI in some sectors. The letter proposes short term and long term research priorities for maximizing the benefit of artificial intelligence to man while attempting to minimize the negative impacts of it on all aspects of society such as labor shortages, gaps in the law dealing with product liability, privacy, safety, and security.

In essence, the AI community, through the letter titled, "Research Priorities for Robust and Beneficial Artificial Intelligence" and other projects like Stanford's *One-Hundred Year study of Artificial Intelligence*, or AI100, acknowledges some degree of threat in the short term. They list threats such as racial, cultural, or economic bias creeping into AI and machine learning. Concerned members of the AI community also consider the long-term existential threats voiced by preeminent scientists such as Stephen Hawking. The renowned computer scientist and artificial intelligence researcher and founder of AI100, Eric Horvitz, wrote in "Reflections and Framing for AI100," "Concerns have been expressed about the possibility of the loss of control of intelligent machines for over a hundred years. The loss of control of AI systems has been a recurrent theme in science fiction and has been raised as a possibility by scientists. Some have speculated that we could one day lose control of AI systems via the rise of superintelligences that do not act in accordance with human wishes—and that such powerful systems would threaten humanity."[22] He goes on to suggest that fears

---

accessed on line March 19, 2017
[22] Eric Horvitz, "Reflections and Framing for AI100, 2014," accessed

about AI need to be addressed by studying the issues and using the results of the studies to shape policy and address growing anxiety. The above examples represent the assertion of self-governance of the AI community. Would it be sufficient for the community of researchers who have organized themselves, written priorities and even set up 100-year studies to protect us from the perceived and real threat posed by the new technology of artificial intelligence?

In "Research Priorities for Robust and Beneficial AI," the authors pose the notion of designing into artificial intelligence a *stop button* designed to stop an outlaw artificial intelligence. Recently, Charles White penned an article titled, "Facebook Robot is Shut Down after It Invented Its Own Language."[23] According to the report, researchers at Facebook shut down two artificial intelligence robots nicknamed Bob and Alice after they abruptly switched from English to a computer language they made up to communicate with each other. "In a Frankenstein twist on their research, the robot abruptly stopped using English and could only be understood by the other AI." In the Facebook case, the off switch was easy to deploy. However, with just a little imagination, it seems possible that a massively intelligent artificial general intelligence would be able to figure out how to disable its stop switch. The artificial intelligence researcher, Oren Etzioni notes that "The A.I. horse has left the barn, and our best bet is to attempt to steer it. A.I. should not be weaponized, and any A.I. must

---

online March 12, 2017
[23] Charles White, "Facebook robot is shut down after it 'invented its own language'," Metro.co.uk, July 31, 2017, accessed online August 2017

have an impregnable 'off switch.'"[24] The authors of "Research Priorities for Robust and Beneficial Artificial Intelligence" note that researchers must consider the situation where a well-designed artificial intelligence that needs to complete a certain task may take actions that are dangerous to man and even other robots. For example, an artificial general intelligence might have the goal of guaranteeing a stable supply of electricity to maintain its own operation and in the pursuit of that goal prevent the scheduled shut down of an aging nuclear power plant to protect its power supply. Researchers need to include in their work a capability to limit the goals of an artificial general intelligence.

Conclusion

Artificial intelligence is not a thing of the future but surrounds us today. Digital assistants like Siri sit at the ready on our smartphones to answer our questions or tell us where the nearest Starbucks is. Supercomputers such as Watson consume vast amounts of information and use it to solve problems not just for *Jeopardy!*, but to help clinicians select cancer treatments and winemakers optimize their water usage. Moreover, artificial intelligence, in many cases, looks at data from a less biased perspective, which offers predictive capabilities that have stunned political pundits and will provide opportunities to profit from election outcomes and market moves. Even though artificial general intelligence remains an aspiration, computer scientists and concerned citizens are organizing and considering

---

[24] Oren Etzione, "How to Regulate Artificial Intelligence," New York Times, September 1, 2017, accessed online September 19, 2017

how we should protect ourselves as we continue to develop more intelligent machines.

Part 2

How to Profit from Artificial Intelligence

Chapter 5
Profitable Applications of AI

Artificial intelligence pervades our world already from the little computers we carry around in our hands in the form of cellphones to the onboard computers that control our cars to the machines moving the stock market with lighting fast stock trading. The potential for profit abounds as the new technology of artificial intelligence transforms every sector of society. The following chapter will take what we have learned about how artificial intelligence works and what it can do and look at ways we can profit commercially, professionally, as a society, and personally during the rapid expansion of artificial intelligence. Identifying pain points in our businesses, society, and personal lives illuminates opportunities to use artificial intelligence and robotics to improve our lives and identify opportunities for profit. Throughout the rest of the book, we will look at both profit and protection standpoints in artificial intelligence and robotics combined in the diverse areas from healthcare to toys to agriculture, and even warfare. For each area, we will apply a *useful test* that serves not only to analyze business opportunities but to understand how advances in AI and robotics will affect our lives. The useful test indicates looking at what we do and finding what makes up the most difficult or painful jobs we have. The test asks you to *find the pain.* Look at where there is pain and see if artificial intelligence and robotics can address that pain. In studying successful entrepreneurs, researchers use the term *pain point* to describe what problem an entrepreneur has solved, and often the success of the business or new product is directly proportional to the size of the problem their product addressed, or in other words, what pain achieved relief.[1] On the other side of the token, you must also identify how robotics combined with AI may bring pain to you, be it personally, at your job or society in general.

Commercial/Professional

As a business owner, employee, entrepreneur, or investor, you need to ask yourself some particular questions. You can ask these questions now that you understand what is happening in the world of machine learning, artificial intelligence, and robotics. Everyone can ask these questions *and should.* The following questions provide examples to stimulate thinking about how artificial intelligence and robotics can impact your work:

---

[1] Jeffrey Carter, "What's a Pain Point?," *Points and Figures*, April 27, 2012, accessed online October 26, 2017

◊ What issues or problems do I have that nag me or rob me of my sleep that artificial intelligence or robotics can address better than I can now?

◊ What thing in my business am I doing poorly, dangerously, or inconsistently that artificial intelligence or robotics would correct?

◊ Do I have (am I) the right people to take advantage of what artificial intelligence or robotics investment will mean for the companies I own or for which I work?

◊ What will be the impact on my business if my competitors use artificial intelligence to gain an advantage over me?

◊ How would artificial intelligence and robotics impact my business?

McKinsey & Co., a leading global management consulting firm, quoted the well-regarded business consultant and writer, Ram Charan in an article titled, "An Executive's Guide to Machine Learning."[2] He was quoted as saying, "Any firm that is not or soon to become a math house is already a legacy company." Such a statement sounds extreme considering some companies are hundreds of years old, making the same product that they have for centuries. Look at Zildjian, the venerable and still dominant maker of cymbals for musicians.[3] Zildjian cymbals began in 1618 when the founder

---

[2] "An Executives Guide to Machine Learning," http://www.mckinsey.com/industries/high-tech/our-insights/an-executives-guide-to-machine-learning, Accessed September, 2016
[3] "Every Company has a History: Ours Starts in 1623," About Zildjian, Zildjian.com, accessed in March, 2016

discovered a new copper alloy that allowed him to make cymbals with a remarkable sound. The emperor of the Ottoman Empire brought him to court and eventually allowed him to form his own private company. The company is still in the family 15 generations later and moved from Turkey to Norwell, MA in the United States. They made cymbals in the jazz era and later in the rock and roll era; it is what they have done for hundreds of years and continue to do so today. However, Zildjian stands as a remarkable exception. In "The Mortality of Companies," the authors analyzed 25,000 publicly traded companies in North America from 1950 to 2009 and found that the average lifespan of a company today sits at ten years compared to 60 years in 1950.[4] Additionally, the lifespan of a company appears to be independent of what segment of the industry they occupy or how established the firm is. The authors note that many companies disappear through mergers and acquisitions and live on in part under other companies' names. Analysts believe that the rise of automation through computerization, including AI and robotics, directly relates to the shortening of company lifespans.[5]

State Street, a major financial firm based in Boston, MA, announced on March 30, 2016, that they would lay off 7,000 employees over the next four years, citing a move toward more automation. *The Boston Globe* reported that State Street intends to become a

---

[4] Madeleine I. G. Daepp, Marcus J. Hamilton, Geoffrey B. West, Luís M. A. Bettencourt, "The mortality of companies," *Journal of the Royal Society*, April 1, 2015, accessed online March 3, 2018
[5] Michael Sheetz, "Technology killing off corporate America: Average life span of companies under 20 years," *CNBC*, August 24, 2017, https://www.cnbc.com/2017/08/24/technology-killing-off-corporations-average-lifespan-of-company-under-20-years.html, accessed online March 3, 2018

more tech-driven company with less reliance on the more traditional manual stock trades handled by people. Although the company announced that turnover would vacate most of the positions or redeployed in new ventures, it is certain that the old jobs will be gone. The nature of work and the types of jobs coming in the near future will look different from those we did in the past, but you can prepare for the future.

I recommend that the reader take out a piece of paper and begin to work on these questions. The answers you provide, although not final, will help you to formulate your interests, drives, fears, and motivations for how you will react to this new and rapidly evolving world of technology and, more specifically, artificial intelligence and robotics. To help answer your questions, first think back to the types of problems artificial intelligence is particularly good at solving. Back in Chapter 1, we looked at different types of machine learning. To get your mind working, we will look at four types of machine learning, the kinds of questions they are best suited for answering, and some examples. The four main categories of artificial intelligence are:

◊ Supervised machine learning for classifying information

◊ Unsupervised machine learning to help find unusual patterns

◊ Reinforcement machine learning for training machines to make decisions

◊ Deep learning helps machines to learn to see, hear, and read for navigation, customer service, and translation

Each of these types of artificial intelligence

alone or in combination with robotics offers new tools to address pain points in our lives commercially, professionally, personally, and as a society.

Think of artificial intelligence as a tool to amplify human intellectual and physical capabilities just as the Industrial Revolution amplified the strength of an individual man. Consider the move from agricultural labor to city centers to jobs in manufacturing, service, and information. With the advent of heavy machinery such as tractors, farms needed fewer hands to work the land. Take for example harvesting corn. One combine can harvest more corn in a minute than a man working all day long. In the 1930s, before combines were available, a farmer could harvest an average of 100 bushels of corn by hand in a nine-hour day. Today's combines can harvest 80 bushels of corn per minute or 100 bushels of corn in under two minutes! One farmer with a Lexion 500 Series Combine can harvest 4,800 bushels of corn in an hour.[6] The point is that significant advances in civilization have come with technological innovations that amplify human power. We are just beginning a new era where computing and robotics will expand human strength and capabilities faster and greater than at any other time in history. The massive change will come with some great winners and some falling farther behind.

Sorting

---

[6] Lexion 500 Series Rotary and Straw Walker Combines, http://www.claasofamerica.com/blueprint/servlet/blob/204870/eee8b0acec56e48bfef1f8a0a61f70fa/2559-lex-l2-500-series-1209-sing-data.pdf, accessed on October 12, 2017

Sorting anything from tomatoes to packages or letters requires the ability to distinguish different things and then decide what to do with them. Single stream recycling refers to the practice of putting all recyclable materials into one bin instead of having residents and businesses sorting recyclables by type such as glass, paper or plastic. Automation promises to help increase the efficiency of sorting recyclables, but single stream sorting causes problems when people put non-recyclables in their bins. One of worst problems is plastic grocery bags. Plastic bags get caught in the mechanical sorting machines, jam them, and need to be cut out by the human workers. Claire Groden wrote in "The American Recycling Business Is a Mess: Can Big Waste Fix It?" that, "Some recycling facilities have to shut down once an hour so that workers can cut layers of plastic bags off the machinery." Down time raises costs.[7]

The challenging task of sorting recyclables offers a new way to combine artificial intelligence and robotics.[8] In 2011, a Finnish company called ZenRobotics introduced an artificial intelligence-driven waste sorter. The recycling sorters were connected to each other to share and learn how to separate different types of trash from the recyclables. Separating clean recyclables from soiled ones poses a big challenge. Apparently, when paper or cardboard such as a used pizza box gets soiled with oil, the paper fibers cannot be separated

[7] Claire Groden, "The American recycling business is a mess: Can Big Waste fix it?," *Fortune,* September 3, 2015
[8] Jennifer Kite-Powel, "This Recycling Robot Uses Artificial Intelligence To Sort Your Recyclables," *Forbes,* Tech #ChangeTheWorld, April 4, 2017, accessed online August, 2017

in the recycling process.[9] To address this issue, a pair of companies from Colorado, AMP Robotics and Alpine Waste & Recycling, developed a robot that sorts food carton waste from other recycling using artificial intelligence that includes pattern recognition to identify food brand names on the labels to aid in sorting out the food containers from other recyclables. Additionally, contamination in recycling dramatically affects the price paid for the material. For example, different types of plastic require separate recycling paths. Bulk Handling Systems offers a new artificial intelligence-driven robotic sorter called Max-AI that can sort six various materials with a very high accuracy rate.[10] Purer bales of recycled plastic significantly increase their market price.

Single stream recycling made recycling very convenient for people by eliminating the need for multiple bins. However, the sorting now needs to be done downstream at material recycling facilities that combine human and mechanical sorting machines. Quality control, sorting out the un-recyclables from the conveyor belt, and un-jamming the sorting machines guarantees a role for humans for the near future, but sorting trash is undoubtedly a pain point suited for artificial intelligence and robotics. In the coming years, garbage sorting robots if they continue to improve and come at a reasonable price, will make the recycling business more profitable.

---

[9] "Frequently Asked Questions: Contamination, Stanford University, PSSI/Stanford Recycling," Stanford.edu, accessed online October, 2017
[10] Mallory Szczepanski, "What Robotics and AI Could Mean for the Future of the Industry (Part One)," *Waste360*, June 13, 2017, accessed online October 14, 2017

Narration/Story Telling

Consolidation in journalism and the continuing reduction in the number of newspapers in the United States has created a shortage of writers and news outlets to cover many local stories. Derek Thompson summed this up in an article in *The Atlantic*, "Since the end of the recession, newspapers, and magazines have shed about 113,000 jobs, while Internet publishing companies have added about 114,000 jobs. That makes it sound as if the jobs are merely shifting from pulp to pixels, but the jobs aren't the same: there is a parallel shift from local news reporting to national news, a result of these sites needing to maximize readership."[11] News organizations are turning to artificial intelligence to write the local stories when there are not enough writers to cover the events. A Chicago-based company called Narrative Science generates sports and business stories using artificial intelligence.[12] At Narrative Science, the computer takes in the statistics of a sporting event and produces a colorful narrative. Additionally, Narrative Science offers a product called Quill that allows a user to tell the computer what the context and intent of the user is and then feed data into the computer to generate a story with supporting *evidence*. For example, if the hometown team lost a game, Quill, knowing that the story is for the local audience, can soften the blow. The text may still sound computer generated, but API claims they can now cover all of Minor League Baseball in the

[11] Derek Thompson, "The Print Apocalypse and How to Survive It," *The Atlantic*, November 3, 2016, accessed online, October 14, 2107
[12] Narrative Science, https://narrativescience.com/#quill-platform

US due to the cost savings from using automation.[13] Additionally, companies such as Automated Insights generate thousands of corporate earnings reports using financial data and artificial intelligence saving the need to hire writers and analysts. Companies with available data such as financial information and the wish to personalize reporting for their clients would benefit from the services of artificial intelligence-driven narratives.

Natural language processing services apply not only to big businesses with big budgets. Small businesses with information about their products, customers, or services such as a realtor with data on housing prices and sales or a trainer with data from gym attendance and fitness trackers can use such information with Automated Insights Wordsmith to generate written descriptions of the information customized to individual clients. Using content generation tools makes possible personalized communications to every client in a matter of seconds rather than hours or days to complete one by one. A fitness trainer could write training summaries, including goals and achievements, to each client in an easy to read summary in a matter of seconds. According to Julia McCoy in an article for the Content Marketing Institute, Automated Insights Wordsmith may fit the budget of some small businesses with an annual contract starting at $2,000 a month.[14]

---

[13] Paul Sawers, "Associated Press expands sports coverage with stories written by machines," *VentureBeat*, July 1, 2016, accessed online October, 2016

[14] Julia McCoy, "Content Creation Robots Are Here [Examples]," *Content Marketing Institute*, November 12, 2017, http://contentmarketinginstitute.com/2017/11/content-creation-robots-examples/, accessed online March 4, 2018

The price may be high, but the cost of hiring a writer or keeping a trainer from training due to his or her writing duties would be significantly higher.

Pattern Detection

Many tasks that require complex pattern recognition have been taken over by computer algorithms because computers can deal with massive amounts of data that no single person would have the time to learn. According to a CNBC report by Luke Graham, cybercrime cost the global economy $450 billion in 2016.[15] Moreover, a good portion of that was credit card fraud. Some jobs, such as fraud detection in the credit card industry, greatly benefit from automation using artificial intelligence. Continuous monitoring of customer behavior, location, and spending trends lends itself perfectly to artificial intelligence especially— unsupervised machine learning and deep learning.[16] Computers carefully compare any transaction made with your credit card against your normal behavior and spending habits. If a customer regularly shops in specific stores or certain online retailers and a transaction occurs outside what the machine considers normal, the transaction halts. The monitoring happens in real-time, and the cardholder may receive a call or message from the credit card company to verify the unusual transaction. The sleepless vigilance of the computer and its machine learning capability to get

---

[15] Luke Graham, "Cybercrime costs the global economy $450 billion: CEO," *CNBC*, February 7, 2017, accessed online October 14, 2017
[16] Ajit Jaokar, "Artificial Intelligence in Fraud Detection," Innovate|Finance Global Summit 2017, March 15, 2017,

to know customer habits easily outstrips the army of analysts that would be needed to understand and monitor the spending habits of all the customers of a global credit card company.

AI Power for Small Businesses

The growth of artificial intelligence over the past ten years depended on massive investments in research and development made by private companies. According to a report from the World Economic Forum in 2017, Microsoft, Google, and Amazon each spent $11.4B, $9.8B, and $9.3B respectively, putting them all in the top ten list of highest investors in technology research and development.[17] The cost of developing technology, including artificial intelligence, goes beyond the reach of many companies. Moreover, most businesses in the United States qualify as small businesses, which means they have fewer than 500 employees. In fact, according to the US Small Business Association Office of Advocacy, 99.7% of all businesses in the US qualify as small businesses, and these small businesses employ 49.2% of all private sector employees in the country.[18] Given the emerging power of artificial intelligence to transform transportation with autonomous cars and power breakthroughs in telecommunications and healthcare, how can a small business use artificial

---

[17] Callum Brodie, "These companies invest the most in research," World Economic Forum, May 15, 2017, accessed online on March 2, 2018
[18] "Frequently Asked Questions about Small Businesses," US Small Business Administration Office of Advocacy, 2017, https://www.sba.gov/sites/default/files/advocacy/SB-FAQ-2017-WEB.pdf?utm_medium=email&utm_source=govdelivery , accessed online March 2, 2018

intelligence to gain an advantage in the market at a price that fits the size and scope of their company?

Small businesses do not need to be tech giants, major corporations, or computer specialists to take advantage of new artificial intelligence-driven tools available today. Successful businesses require customers to purchase their goods and services, and managing customer relationships forms a crucial part of business success. Since the beginning of business, success depends on knowing customer needs and making sales, which often happens with sales calls or marketing campaigns. Before computers and the internet, keeping track of customer relationships depended on paper records and the Rolodex (a rotating spindle of cards containing names, addresses, and phone numbers). In the 1980s, software companies began developing computerized systems to replace the Rolodex and provide companies with additional insights to improve sales effectiveness. The systems known as CRMs for customer relationship management began to evolve through the 1990s into sophisticated tools to help manage customers, recommend the best clients to call on a given day, and drive advertising decisions. A leading company providing software for customer relationship management called Salesforce now offers a tool known as Einstein Analytics as an additional service to complement their traditional CRM tools. According to Salesforce, Einstein Analytics uses artificial intelligence to gain insights and make sales and marketing recommendations and predictions based on billions of pieces of information. According to Salesforce.com, their basic CRM solution for small businesses starts at $25/month/user with additional costs Einstein analytics unspecified on their website.[19] However, Infotechlead.com indicated in an article titled "Salesforce Reveals the Price of Einstein Analytics," that the additional Einstein components cost $75 to $150 per month extra.[20]

Advertising allows businesses to reach potential customers with the intent to drive sales. Marketers have always tried to reach target audiences by tailoring advertisements to potential customers, which is why the advertising during a football game looks much different from the advertising during a television morning show. Television advertising has always been expensive, but with the massive expansion of internet, or e-commerce, small businesses can now reach an unprecedented global pool of customers at much lower costs than television advertising. With internet marketing, small businesses have global reach, but they must compete with other companies to emerge at the top of the search results on Google or Bing. Google offers AdWords to support online advertising. AdWords works as a pay-per-click, or PPC service, that charges the advertiser a fee every time someone clicks on an advertisement that came up during a Google search. AdWords uses sophisticated artificial intelligence called Life Event Targeting to point ads to target audiences. For example, if a wedding planner advertises with AdWords, Google does not just match searches that include wedding and planner. Instead, Google's AI in Life Event Targeting

[19] https://www.salesforce.com/solutions/small-business-solutions/pricing/sales/?d=70130000002DsrA, accessed on March 2, 2018

[20] "Salesforce reveals the price of Einstein Analytics," *Infotechlead*, June 16, 2017, http://www.infotechlead.com/analytics/salesforce-reveals-the-price-of-einstein-analytics-49711, accessed online March 3, 2018

learns that people approaching a big event like a wedding, graduation, new baby, or new home purchase will behave differently online before the events occur.[21] AdWords through Life Event Targeting can know even before you tell any of your friends and family that you are pregnant or getting a divorce. With such knowledge, AdWords can specifically target individuals with very tailored advertising. The price of AdWords follows a formula that depends on the terms used and the quality of the advertisement. According to a blog post titled, "How Much Does Google AdWords Cost?" the price per click from Google AdWords ranges from $1 to $50 per click.[22] Although not inexpensive, AdWords advertising offers an alternative to traditional print and media advertising. Careful use of keywords and a good advertisement makes use of artificial intelligence for very targeted small business advertising.

Great strides in developing artificial intelligence have depended in part on massive investments in research and development by technology giants such as Microsoft and Google. However, in the United States, over 99% of all businesses qualify as small businesses. Most small businesses will not have the budget or expertise to develop artificial intelligence designed to help their business grow and gain strength in the market, but that does not mean that small businesses cannot affordably use artificial intelligence today.

---

[21] Raghav Haran, "Google Marketing Next: AI Will Be a Major Feature of AdWords in 2017," *Singlegrain*, https://www.singlegrain.com/news/google-marketing-next-ai-will-major-feature-adwords-2017/, accessed online March 1, 2018

[22] Dan Shewan, "How Much Does Google AdWords Cost?," WordStream, December 11, 2017, https://www.wordstream.com/blog/ws/2015/05/21/how-much-does-adwords-cost, accessed online March 1, 2018

Customer relationship software such as Salesforce's Einstein Analytics and advertising services such as Google's AdWords with Life Event Targeting bring the power of artificial intelligence to better manage sales and marketing to get an edge in the market.

Prediction

Human resources (HR) departments play a useful role in many businesses. HR is responsible for activities such as hiring personnel, training, benefits, administration, providing a consistent work environment, and even serving as a neutral third party in disputes among employees. Hiring requires significant time to review tens, even hundreds, of candidates for open positions, and HR often acts as the frontline, filtering the candidate list down to a reasonably short list for review by the manager who posted the job. Given the fact that many people comfortably use online, AI-driven matchmaking services to find a date or a mate, HR may be equally comfortable letting one of the growing numbers of AI-driven job candidate selection systems to do the job for them. Back in the 1990s, electronic job posting services like Monster. com and CareerBuilder.com massively expanded the reach of employers to offer jobs to thousands of potential employees online and receive applications at an unprecedented number. The massive expansion of reachable job candidates created a problem that HR did not have to face before the internet—too many applications to read. Recently, the job site Indeed. com claimed on their blog that they exceeded 200 million unique visitors and hosted over 16 million job listings.[23] Given the volume and reach of employers and job seekers, new AI technology continues to develop products that help recruiters and HR departments better cope with so many choices. AI helps to sort and evaluate hundreds of applications, recommending top candidates for the human recruiters to assess. Good AI systems will also keep track of the success of hires as they mature in the company, learning what different metrics define high performing employees and use this information to select candidates in the future.

Just as AI tools are being designed to assist HR departments in sorting candidates to fill open positions, some tools like Mya[24] provide AI services also to help early-career job candidates find positions. The AI known as Mya automates much of the recruiting process. Mya asks questions of candidates and, with natural language processing, learns particulars about each candidate, providing insights and real-time feedback on common questions. Mya functions as the first layer of candidate screening to identify best candidates even before a job application gets filled out.

Beyond hiring, AI finds increasing deployment in the areas of employee retention and surveillance. The surveillance services often promise improved productivity, morale, quality of life, or security. Veriato Inc. offers an employee surveillance platform called Variato 360 that provides visibility to employers into the online activity of all their workers. Not only does the platform open a window onto what their employees do online while at work, but Variato also provides analytics that helps managers to evaluate good activity from bad activity. The system collects periodic screenshots of an employee's internet browsing as evidence that can be played back for investigation and support for allegations of misbehavior. Just seeing what an employee or contractor is doing on their computer may not tell a compelling story; knowing the sentiment of the employee's communications also may help to identify different issues such as job dissatisfaction or conflict with management. Veriato analyzes email messages for content and sentiment that would indicate if an employee is no longer satisfied with their work and may need more careful coaching or monitoring.

Sapience Analytics Inc. is a people analytics company that provides tools to monitor workers activity on their computers. Sapience Analytics' homepage leads with the phrase, "End distractions. Enter mindfulness."[25] The company offers the notion that procrastination not only robs companies of productivity but also suggests that procrastination implies that someone is stressed out. They claim that by displaying to employees and managers alike how and where employees spend their work time online reduces their meandering on the internet. Sapience suggests that awareness of one's habits like frequent checking of Facebook on the company clock helps employees to be more focused or *mindful* of their habits, which subsequently improves performance. Sapience even offers a version of their product for home use that helps people stay aware of how they are using their time online in comparison with the time they are spending with their family or on

---

[25] www.sapience.net

other activities. Claims that mindfulness will enhance productivity, however, do not stand on strong scientific research. In fact, the clinical neuroscientist Nicholas Van Dam, PhD and others in their research article titled, "Mind the Hype: A Critical Evaluation and Prescriptive Agenda for Research on Mindfulness and Meditation" caution that little quality research exists on the effects of mindfulness on behavior.[26] They write, "Misinformation and poor methodology associated with past studies of mindfulness may lead public consumers to be harmed, misled, and disappointed." Be careful of accepting claims that being aware of how one spends their time will lead to achieving one's goals and expectations.

Improve Sales

Businesses face continuous competition from companies in the same sector and the threat of market changes that will make their business obsolete. There are many cited examples of companies that did not see their competition coming or the shift in technology before it was too late. Smith Corona led the world of mechanical typewriters manufacturing for almost 100 years, innovating in the typewriter space first with portable mechanical typewriters, after World War II emerging as a leader in the electric typewriter, and holding a strong position in word processing in the 1980s.[27] Smith Corona did not see its competition

---

[26] Nicholas T. Van Dam et al, "Mind the Hype: A Critical Evaluation and Prescriptive Agenda for Research on Mindfulness and Meditation," *Perspectives on Psychological Science*, 13, 36-61, 2018

[27] Smith Corona History, https://www.smithcorona.com/history.html, accessed online October 15, 2017

coming, and the personal computer eliminated the need for mechanical typewriters and word processors. Smith Corona exists today in the business of thermal labels but does not hold the global position it once did. In contrast, some companies today maintain market share in an area completely different from where they started. For example, few may know that the Suzuki Corporation before it became a global name in marine, automotive, and performance motorcycle manufacturing, began in 1909 as a loom manufacturer in Japan founded by Michio Suzuki.[28] After more than forty years in the loom business, Suzuki introduced a small clip-on motor for bicycles, thus entering the motorcycle field and today producing world champion winning racing motorcycles. Similarly, the world's leading video game maker, Nintendo, did not start back in 1889 as a maker of electronics and video games.[29] Nintendo for over seventy years was a manufacturer of trading cards. During the 1960s, Nintendo diversified into different areas, including a taxi business and a chain of short stay *love hotels*. Transitioning to electronic toys and, eventually, video games moved Nintendo to the forefront of electronic gaming.

Businesses exist in competitive environments, and business leaders need information about their processes, competition, and changing markets to know when to act and even pivot to new markets. Artificial intelligence increasingly provides insights and tools to improve performance and insights into to how competitors and

[28] History, Suzuki Corporate website, http://www.globalsuzuki.com/corporate/history/index.html, accessed online October 15, 2017
[29] Florent Georges and Isao Yamazaki, *The History of Nintendo 1889-1980*, Les Editions Pix'N Love, 2012

markets are moving. Earlier in Chapter 2, we looked at how a research team at Microsoft Azure used artificial intelligence to predict customer churn in the highly competitive cellular telephone market. Netflix, the video streaming giant, now has customers in over 190 countries around the world and is deploying artificial intelligence to provide a higher quality streaming experience. Citing an interview on BBC conducted by Francine Stock with the CEO of Netflix in early 2017, an article titled, "Netflix Is Using AI to Conquer the World... and Bandwidth Issues," describes how Netflix is dealing with bandwidth. Netflix must tackle the problem of low bandwidth (low signal, so the picture gets lost) in some of the developing markets that do not have very robust internet service.[30] Netflix uses artificial intelligence with training by humans to analyze video pictures for quality. The artificial intelligence learns which parts of the frames of the video it can compress without losing picture quality. Video compression, in other words, figures out things like a background that does not change much and can reuse the background over again in the next frames of the video. Such compression means that less information needs to be sent over the internet thereby improving picture quality in markets with slower internet speeds. Knowing that competition for video streaming will only increase in coming years across the globe, Netflix is using artificial intelligence to stay ahead of its competitors.

---

[30] Danny Vena, "Netflix Is Using AI to Conquer the World... and Bandwidth Issues," *The Motley Fool*, March 21, 2017, accessed online October 16, 2107

Improve Efficiency

Efficiency in business drives shorter production times, lower costs, and higher profits. Artificial intelligence already plays a role in driving higher efficiency in companies around the world. There are many ways that learning machines with their capabilities in sorting, pattern recognition, prediction, and decision making can help businesses. From the production line to customer service and sales, artificial intelligence has many applications in improving efficiency. Recently, in a joint analysis between the MIT Sloan School of Management and Boston Consulting Group titled, "Reshaping Business with Artificial Intelligence," the authors noted that Airbus recognized that production inefficiencies in the production of their new A350 aircraft threatened to cost the company billions of euros.[31] The competitor for the A350 is the Boeing 787 Dreamliner, and according to Boeing the 787 has over 2.3 million parts, ranging from *fasten seatbelt* signs to jet engines.[32] The massive number of parts that go into a modern composite jetliner serves to emphasize the complexity of the production line. Airbus built an artificial intelligence system that recommends on the spot a solution to production problems. "It combined data from past production programs, continuing input from the A350 program, fuzzy matching, and a self-

---

[31] Sam Ransbotham, David Kiron, Philipp Gerbert, and Martin Reeves, "Reshaping Business with Artificial Intelligence," *MIT Sloan Management Review,* September 6, 2017, Accessed online October 16, 2017

[32] http://787updates.newairplane.com/787-Suppliers/World-Class-Supplier-Quality

learning algorithm to identify patterns in production problems. In some areas, the system matches about 70% of the production disruptions to solutions used previously — in near real time."[33] Using artificial intelligence, Airbus can capture and re-use or create new solutions to production problems and feed them to the production line in a continuous manner.

Customer service forms an essential piece of most successful companies. Customer service in the insurance industry is no exception. In the insurance industry, *efficiency*, the time to process insurance claims and settle disputes, directly relates to customer satisfaction and the bottom line. The Chinese insurance company, Ping An Insurance Group, recently rolled out new artificial intelligence systems to make their customer service more efficient.[34] One product uses pattern recognition in photographs of automobile accidents to identify the automobile's make and model, read the license plate, and estimate the cost of repairs almost instantly, saving the time of an assessor to come out to the accident scene and subsequently write up the assessment. According to James Garner, the Chief Strategist at Ping An, Ping An invests over $1 billion per year in technology research, ranking the 14th of all companies globally in technology research spending, and that their facial recognition capabilities are the best in the business.[35] The company uses artificial

---

[33] Sam Ransbotham, David Kiron, Philipp Gerbert, and Martin Reeves, "Reshaping Business with Artificial Intelligence," *MIT Sloan Management Review,* September 6, 2017, Accessed online October 16, 2017
[34] Chen Meiling, "Major insurer Ping An using AI to improve efficiency," chinadaily.com.cn,: September 6, 2017, accessed online October, 18, 2017
[35] Ping An is far ahead in using A.I. in the insurance business: exec,

intelligence for facial recognition and voice signatures to create biological files on all their clients in order to improve customer service, claiming that since they have used artificial intelligence, it has reduced the time to resolve claims to 30 minutes instead of 3 days as before. Artificial intelligence undoubtedly can increase efficiency.

Prediction

Who has not wished they could see the future? People have been looking to oracles, palm readers, divination, and even statistics to help them see what will happen. Questions such as will the stock market go up, can I trust someone will pay me back, will this person be right for the job, how will the harvest be this year, what are our chances in battle tomorrow, will my product sell at this price, and many, many more have occupied the thoughts of everyone from business leaders and government officials to soldiers and civilians for millennia. Making the right prediction can bring wealth or success, just look at the consistent winners in the field of investing such as Warren Buffet, known as the Oracle of Omaha, who has become one of the wealthiest people in the world through investments he has made throughout his life. On the other side of the coin, many a fortune has been lost on making the wrong bet. Take for example the Hunt brothers, Texas billionaires who lost it all trying to corner the

James Garner, CNBC, 8:18 PM ET Thu, 17 Aug 2017, https://www.cnbc.com/video/2017/08/17/ping-an-is-far-ahead-in-using-a-i-in-the-insurance-business-exec.html, accessed online October 18, 2017 silver market in the 1980s.[36] Prediction has power and plays a role in significant decision making but also in functioning in our everyday lives. When you drive a car, you are always making predictions about how the other drivers will behave, what is on the road up ahead, and more. Artificial intelligence with more data, faster computers, and more memory continues to improve at making predictions and is being deployed in many beneficial ways today.

Many firms that rely on predictions benefit from artificial intelligence because computers can deal with massive amounts of data. Tasks such as medical diagnostics, fraud detection, data security, and identifying market trends benefit massively from non-biased algorithms. On June 23, 2016, the United Kingdom in a historic national referendum chose to leave the European Union. The vote stunned Europe and London and shook the financial markets, plunging the Pound Sterling to a 30-year low against the dollar. Global stock markets tumbled too with steep drops across Europe from the FTSE250 down 12% to Spain's IBEX down 11%. In a single day, trillions of dollars of value were erased from the global market based on the UK leaving the EU. Market turmoil sees many losers but also offers an opportunity for significant gains for the shrewd investor. According to authors in *The Wall Street Journal*, the machines did better than people in making profits from the market falling.[37] Although the

---

[36] Ben Christopher, "How the Hunt Brothers Cornered the Silver Market and Then Lost It All," *Priceonomics*, Aug 4, 2016, accessed online October 22, 2017
[37] "Who Made Money in the Brexit Chaos? Machines, Not Humans," *The Wall Street Journal*, June 29, 2016, The *Wall Street Journal*, July 4, 2016

polls leading up to the vote were too close to call, many investors were biased toward the outcome they wanted to see—Britain staying in the EU. Investors do not like uncertainty, and Britain remaining in the EU as it had been for decades was a much more certain future. However, the machines in some cases told a different, unbiased story, and the investors who stuck with their computer models turned profits in the down market.

Over many years, the analysts and brokers in all the great financial markets such as the New York Stock Exchange and the London Stock Exchange have increasingly employed mathematics and automation to gain advantages in the markets with better predictions and faster trades. In an article from *The Wall Street Journal* titled, "The Quants Run the Show Now," the authors, Geoffrey Zuckerman and Bradley Hope, describe how *quants*, or quantitative analysts, use large data sets, mathematics, and artificial intelligence to power their trading. "Up and down Wall Street, algorithmic-driven trading and the quants who use sophisticated statistical models to find attractive trades are taking over the investment world."[38] Moreover, financial firms are hiring the quantitative analysts to develop algorithms, but some firms realize that they need traditional human analysts to work alongside the quants to make the best trades, combining the skills and information from both. Artificial intelligence can help to make sense of the reams of data coming from multiple sources such as the dark web, hotel bookings, and online consumer sentiment by digesting and using this

---

[38] Geoffrey Zuckerman and Bradley Hope, "The Quants Run the Show Now," *The Wall Street Journal*, May 21, 2017, accessed online September 20, 2017

information to inform predictions; however, people have their own way of finding undervalued assets. Recently, the financial giant, JP Morgan Chase, announced that it would deploy the first artificial intelligence-based trader, employing deep learning and reinforcement learning to execute large trades with the goal of not destabilizing prices.[39] The artificial intelligence known as LOXM does not decide what to sell but how to sell. People working alongside artificial intelligence, using the strength of artificial intelligence and human skills of judgment and intuition, is referred to as *centaurs* after the mythical half-man half-horse creatures in mythology. As one analyst pointed out in "The Quants Run the Show Now," many of the algorithms may become similar, limiting the competitive advantage and possibly destabilizing markets, which says that human traders thinking creatively will emerge again as critical to succeeding in the market. In short, identifying good centaur mixes will bring the ultimate advantage. Mark Stefik, a research fellow at Palo Alto Research Center Incorporated (PARC), adds, "Even with today's powerful and deep machine learning approaches, computers typically react poorly in situations different from those where they have been trained. People draw on a lifetime of experience and their common sense and bring that to the centaur team."[40] *Centaurs*, or human-machine teams, mostly exist as research projects and evolving systems for large corporations and the military, but applications for the individuals will also emerge soon to assist people faced with large complex sets of

---

[39] Laura Noonan, "JPMorgan develops robot to execute trades," *Financial Times*, July 31, 2017, accessed online October 12, 2017
[40] Mark Stefik, "Half-Human, Half-Computer? Meet the Modern

data such as a radiologist facing many types of imaging such as MRIs and CT Scans.

Security

Security involves protecting people and property from harm or theft. Improved security protects profits and prevents losses. With more sophisticated threats such as identity theft, computer viruses, and cyberattacks adding to the traditional physical treats such as theft and assault, security remains a major concern for people today. Artificial intelligence will be a powerful tool in security both in the physical and digital worlds. Roman V. Yampolskiy, an associate professor at the Speed School of Engineering, University of Louisville, wrote in the *Harvard Business Review* that artificial intelligence now provides criminals and enemy states powerful tools to attack governments, businesses, and individual accounts.[41] On the other hand, he sees that we can use artificial intelligence as a defense against malevolent artificial intelligence. He adds, "Business leaders are advised to familiarize themselves with the cutting edge of AI safety and security research, which at the moment is sadly similar to the state of cybersecurity in the 1990s, and [...] Hiring a dedicated AI safety expert may be an important next step, as most cybersecurity experts are not trained in anticipating or preventing attacks against intelligent systems."

Security most certainly benefits from machine

Centaur," PARC blog, January 25, 2017, accessed online March 4, 2018
[41] Roman V. Yampolskiy, "AI Is the Future of Cybersecurity, for Better and for Worse," *Harvard Business Review*, May 8, 2017, accessed online August 20, 2017

learning beyond fraud detection, with significant advances in facial recognition. Back in the 1960s, research in computer facial recognition began with using specific coordinates determined by a person to help the computer compare against the coordinates of facial features derived from a picture such as the distance between the eyes. Better algorithms, including machine learning, have significantly improved facial recognition software. In fact, even your little phone and digital cameras use facial recognition to help identify faces to assist in adjusting focus for better pictures. Security systems continue to consume the latest in facial recognition technology to search for the faces of criminals and cheats in crowds such as those entering a casino or airport. Companies such as Imagis Technologies, Inc. supply facial recognition systems to law enforcement that can be run right from a police cruiser via a cellular connection. Fewer security guards are needed to keep an eye out for known criminals or suspected card counters at casinos, and direct access to databases with millions of faces massively expands the reach of law enforcement.

Immigration checkpoints are even becoming automated. Imagine walking through passport check without waiting in line to have the terse conversation with the border control guard about where you have been, how long you were away, and what the purpose of your trip was. New Zealand and Australia now use SmartGate technology that allows their citizens traveling from select countries such as the United Kingdom and the United States to self-process through customs and immigration. The electronic passport has a chip containing essential information, including a digital photo of the passport holder, allowing a facial recognition processing program to confirm that the face in the passport matches the person walking through and that the picture matches the one in the central database. Self-service customs and immigration will no doubt expand. Consider the proliferation of self-service check-in kiosks at all the airline desks at all the airports. AI will continue to change how security at airports and other public places work.

Large countries have massive borders. Consider that the border between Mexico and the United States is over 1,900 miles long and is densely populated in some places and nearly unpopulated in others, stretching through harsh, desert, and mountainous terrain. Patrolling such a long border poses many challenges for border patrol agents. Young-Jun Son, professor and head of the University of Arizona Department of Systems and Industrial Engineering, received a $750,000 grant from the US Airforce to develop artificial intelligence systems that will use data from remote sensors such as cameras on the ground as well as information from airborne and ground-based autonomous vehicles. Compiling all this information, the artificial intelligence will better help deploy human patrol agents. "Using NASA geographical data from the border, the UA researchers have written hundreds of algorithms to simulate and predict how groups of people may move when traveling on flat desert and mountains, uninhabited areas and cities, in dry, dusty conditions or during monsoons."[42]

Another excellent example of using artificial intelligence in security is in wildlife protection. In

---

[42] "Smarter control for border patrol," *ScienceDaily*. July 7, 2017, accessed online August 20, 2017

2013, researchers at the University of Southern California developed a system that uses AI and a type of mathematics called *game theory* to protect endangered species such as the African elephant and the rhinoceros.[43]   They named the program PAWS, which stands for Protection Assistant for Wildlife Security. PAWS helps park rangers to optimally select the best anti-poaching patrol routes to protect the most animals and intercept poachers. The tool uses artificial intelligence to analyze features of the terrain such as hills, mountains, rivers, lakes, and forests as well as known migration paths for certain animals. The system learns over time from the results of every patrol and uses game theory, a mathematical way of guessing what your opponent will do, to predict where the poachers will be. The PAWS system has given the rangers a much better chance to protect the animals in their park. With tens of thousands of animals being poached each year for their skins, medicinal qualities such as the rhino horn for impotence, or hunting for sport, PAWS provides a great example of artificial intelligence enhancing security capabilities.

One of the drawbacks of PAWS stems from the fact that it makes the route calculations based on past route performance.   Another artificial intelligence system developed by a research team at Deakin University in Australia employs real-time data about the location of the protected animals in the park.[44] Using motion-activated cameras deployed throughout the park, the

---

[43] Jackie Snow, "Rangers Use Artificial Intelligence to Fight Poachers," *National Geographic*, June 12, 2016, accessed online July 14, 2017
[44] Yong Xiang, "How artificial intelligence is revolutionizing conservation," Deakin University, Accessed online July 14, 2017

system collects information about animal whereabouts. Monitoring and analyzing so much video footage poses a big problem for the park rangers who do not have enough staff to watch all the cameras all the time. But the artificial intelligence learns each kind of animal and then continuously and tirelessly keeps track of where all the animals are and even predicting where they will go next. Knowing where all the animals are as well as tracking poachers and their vehicles, the system can help the rangers better use their limited time and resources.

For individuals or small businesses, enhanced security employing artificial intelligence hit the market in 2016 for home use in the form of outdoor security cameras from Nest. Nest offers cameras that use Google artificial intelligence for face recognition to determine if the thing approaching the house or business is a human or not. Nest Cam Outdoors retails for $199 each and notifies the user on their smartphone when a human comes into the frame of the camera. The user can then review the footage and even speak to the intruder from the phone through speakers in the camera. In an article on Security Info Watch titled, "Future of Residential Security Tech on Display at CES 2017," the author describes advances in home security that use artificial intelligence to take information from sensors in the home such as motion detectors and devices like smart locks to learn the normal comings and goings of a home.[45] People in a home will, for example, let the dog out at a regular times or open the garage at a certain

---

[45] Joel Griffin, "Future of residential security tech on display at CES 2017," Security Info Watch, January 6, 2017, accessed online March 4, 2018

time to go to work each day. Using these patterns, the system can alert the owner when a door opens or a car pulls into the driveway at an odd time. The security company Lock.com also plans to launch unmanned aerial vehicles, or drones, that will investigate suspicious activity around a house or business. The video feed will be available by smartphone or computer and sharable with law enforcement. Personal and small business security serves to protect life and profits through deterring theft and unauthorized access.

Transforming Products

Products go through transformations as new technologies become available and mass producible. Before the availability of cheap transistors and other electronic parts, toys did not have electricity to light up, talk, or move around; instead, they had mechanical features such as a windup mechanism that turns wheels or makes sounds. Products today are increasingly using artificial intelligence to improve their utility and appeal. I remember the exciting moment as a young boy when the little electric slot cars that we would race around the track for hours became available with working head lights so we could race in the dark. Such a simple addition of technology added a whole new dimension to AFX racing. The applications of artificial intelligence to products opens the door to new ways of people interacting with their things than ever before.

Toy makers continue to expand the use of AI and robotics in toys such as Cognitoys' Dino. Dino looks like a friendly green or pink plastic dinosaur, which, through a WiFi enabled connection with IBM's Watson, interacts with and learns about a child.

Dino even adapts to a child's development level and learning style to be a companion and instructor. The goal of Cognitoys is for Dino to become your child's companion and to use what Watson learns to help a child to navigate the world. There is more according to the manufacturer's website, "Dinos don't just respond, they respond intelligently. If a child's scared, the Dino consoles the child and encourages them to speak to an adult they trust. Or, if a child tells the Dino they're sad, the toy may suggest meditation or listening to a funny joke."[46] Many more toymakers are adopting and developing AI and robotics to attract the attention of parents and children today.

Mattel Corporation stands as the largest toy manufacturer in the world (based on revenue) and provides many toys for children including Matchbox Cars, Fisher Price toys, and the truly iconic Barbie doll. Since Barbie's introduction in 1959, the doll has been at the center of many controversies and simultaneously a beloved toy for millions of girls around the world. According to Mattel, over one billion Barbie dolls have been sold worldwide, making Barbie by far the most popular fashion doll of all time. As a fashion doll, Barbie has also drawn criticism for creating unrealistic expectations of beauty and body image for young girls. In response to the criticisms following Barbie, Mattel continues to innovate and improve the doll. The fact page on Barbiemedia.com reports that Barbie has had over 180 inspirational careers and that over 50 fashion designers have designed for her. In addition to Barbie's appearance, Mattel has sought to innovate Barbie

---

[46] How it works, Cognitoys website, https://cognitoys.com/pages/about, accessed October 24, 2017

technically over the years. Following Mattel's launch of the world's first talking doll, *Chatty Cathy*, Barbie was given a voice in 1968 that was activated by a pull string at the back of her neck. Advances in manufacturing and technology also contribute to Barbie's evolution, including a WiFi-enabled Barbie introduced in 2015 called *Hello Barbie*.

Hello Barbie employs many advanced technologies to allow her to talk, listen, and carry on a conversation with young girls. Hello Barbie uses artificial intelligence for voice recognition, machine learning to tailor conversations, and cloud computing to process and store all the information needed to personalize each person's experience with Barbie. In 2017, it was announced that Barbie will take another significant step further into the digital world when Mattel said it would launch a new holographic Barbie for the Christmas rush (Editor's Note: As of the time of the publication of this book, Mattel has not yet released the Hello Barbie hologram, and it will be delayed until 2018[47]). For the first time, Barbie will leave the physical plane and be available as an interactive, dancing, and talking three-dimensional hologram. Instead of the famous 1/6 scale plastic doll that girls can play with and share with their friends, holographic Barbie will live in a clear plastic box atop a base that sparkles with LED lights and emblazoned with the name *Barbie*. Check out some of the early prototypes of the playset on YouTube to see some remarkable technology that will make Barbie the

[47] Miller, Andrea. "Mattel Delays Kids' Voice Assistant Hello Barbie Hologram until 2018." *ABC News*, ABC News Network, 6 Nov. 2017, abcnews.go.com/Technology/mattel-delays-kids-voice-assistant-barbie-hologram-2018/story?id=50963466.

first, "holographic, digital friend for little girls."[48] The prototype shows a moving, dancing, and configurable Barbie that will know a girl's name, make conversation, and even remember past conversations to act truly like a friend. The hologram can also be customized with a simple command of, "Hello, Barbie. Will you change my Barbie?" The Barbie hologram will then change its ethnicity, body type, and clothes. Also, Barbie is connected to information available on the web. So, like Amazon Echo or Google Dot, you can ask for a weather report, and Barbie will tell you the weather and even recommend what is appropriate to wear.

Since holographic Barbie can dance and hold conversations but now has lost all physicality, is it still a toy in the traditional sense? When I was little, I remember my sisters and their friends playing with Barbies for hours, even days on end, building elaborate scenes with houses, animals, and cars. I did not take part, but I could hear them immersed in play with unending scenarios and conversations. Toys as simple as a few wooden blocks or dolls made of plastic or cloth function as a conduit to a limitless world of imagination. With Barbie going digital, will the Barbie digital friend still have the capacity for such endless play? It remains to be seen once holographic Barbie is released to the public if the users will see her as a new piece of tech or an actual toy. Holographic Barbie may usher in a new era of digital toys, or it may represent the departure of Barbie from the realm of toys and into a digital AI persona. With advances in robotics combined with AI, Barbie may be more interactive than ever before.

---

[48] YouTube of Holographic Barbie https://youtu.be/GlQBjsxE0CY

Toymakers continuously struggle to innovate to drive new sales and to make toys that appeal to the next generation of children.

The above descriptions of product innovation serve as examples of integrating new technologies into existing products to drive sales and profits. Businesses of all types should consider how artificial intelligence or robotics can transform their current products to anticipate market needs and increase profits. Adding a robot such as Pepper from SoftBank that uses artificial intelligence to greet customers may draw business and foot traffic. Although not inexpensive at $1,700 for the robot with a $223 per month fee that includes maintenance and insurance, Pepper can interact with customers in a friendly and novel way.[49] Existing businesses such as fitness and sports trainers can also take advantage of powerful applications to enhance their existing training techniques. For example, a golf or tennis instructor or baseball coach may use an inexpensive product called Zepp 3D Swing Analyzer ($95.88 at Walmart) that works with a sensor, a smartphone, and artificial intelligence to analyze swing, make meaningful suggestions for improvement, and even show comparisons with professional athletes. Such a tool as Zepp 3D Swing may help a coach or instructor improve their ability to help athletes and new customers reach to improve their game in a novel way.

Customer Service

---

[49] Larissa Faw, "Pepper The Robot Will Be Your Companion (For A Price), MediaPost, May 22, 2016, accessed online March 7, 2018

The vast world of customer service covers everything from retail assistance to product support. Many of these areas have been impacted already by artificial intelligence, especially with the growing sophistication of chatbots that can understand spoken language and help customers solve their problems, whether it is a product that is not working or making an insurance claim. As mentioned before, some insurance companies such as Ping An in China are using artificial intelligence to significantly speed up the process of resolving insurance claims. Most of the major telecommunications and financial services use an artificial intelligence-based phone support system to be the first encounter with customers. Other types of customer service are using artificial intelligence to support human skills in the retail sector as well as at home with systems like Amazon's Alexa or Google's Pixel.

Who has not struggled to find clothes that best fit one's style, budget, and unique measurements? It may not seem like an initial fit, but the clothing industry has begun to embrace artificial intelligence to provide personal style assistants and even custom clothes fitted to one's exact measurements. The top clothing companies according to an article titled, "AI in Fashion—Present and Future Applications," have primarily invested in artificial intelligence in the area of chatbots to improve customer experience and to provide a sort of automated stylist.[50] Some companies have gone further to employ artificial intelligence to work with their stylists and manufacturers to usher in a new

[50] Kumba Senaar, "AI in Fashion – Present and Future Applications," Techemergence, August 14, 2017 accessed online October 24, 2017

era in the clothing industry. In Mumbai, India, a father and son team have developed a business that will use artificial intelligence combined with expert tailors to provide custom fit shirts made of high-grade Egyptian cotton. The company called Crisp Clothing claims that, "By simply referencing your height & weight, we can accurately predict your body measurements in under a minute, using advanced artificial intelligence and machine learning algorithms. No measuring tape required. True 21st-century tailoring."[51] Another company called Stitch Fix claims, "Stitch Fix is the world's leading online personal styling service that blends the art of expert personal styling with the science of algorithms to deliver apparel and accessories tailored to your taste, budget, and lifestyle."[52] With sales of over $730 million and 5,700 employees, Stich Fix has taken great advantage of artificial intelligence combined with human expertise to provide a fashion subscription service that continues to grow. In an article from the *Harvard Business Review* titled, "How One Company Blends AI and Human Expertise," the authors point out that, "Stitch Fix's approach illustrates three lessons about how to combine human expertise with AI systems. First, it's important to keep humans in the business-process loop; machines can't do it alone. Second, companies can use machines to supercharge the productivity and effectiveness of workers in

---

[51] Perfect Fitting Custom Shirt by Crisp Clothing, Kickstarter description, https://www.kickstarter.com/projects/crispclothing/the-perfect-fitting-custom-shirt-using-3d-measurin?ref=pr.go2.fund&utm_medium=referral&utm_source=pr.go2.fund, July 2017, accessed online October 24, 2017

[52] "Stitch Fix Enjoys Growth & Client Loyalty," Stitch Fix Corporate press release, May 10, 2017, accessed online October 24, 2017

unprecedented ways. And third, various machine-learning techniques should be combined to effectively identify insights and foster innovation."[53]

Stitch Fix uses different types of artificial intelligence to assist the stylists they employ who mostly work from home. The artificial intelligence makes suggestions to the stylists who assemble collections to ship to their clients. Making the right recommendations is essential to the success of Stitch Fix, and their researchers are using more and more information such as computer analysis with computer vision to identify style trends, learn from customer satisfaction, and make recommendations based on customer similarity. Stitch Fix is a great example of a *centaur system*, or a human-machine team that takes the best from machine learning combined with human judgment, to open a new sector of jobs in the new information age.

Creativity

Creativity stands as one of the essential elements of human consciousness and intelligence. Examples of human creativity surround us and enhance our lives from simple things like the fork and knife for dining to history-making creativity such as the printing press and calculus. As we look at the creative outputs of computers today, they have not contributed *H-creativity*, or creating something that has not been thought of before, that will affect an entire society like Isaac Newton's calculus, but we may be seeing new smaller

---

[53] H. James Wilson, Paul Daugherty, and Prashant Shukla, "How One Company Blends AI and Human Expertise," *The Harvard Business Review*, November 21, 2016

sources of creativity beyond man. AlphaGo, Google's Go-playing AI program, created a new way to play Go. The stunning victory over Lee Se-dol watched by over 60 million Chinese fans supports the notion that computers indeed can exhibit a sort of creativity, but it remains to be seen just *how* creative AI really is.

Artificial intelligence continues to demonstrate problem-solving skills in other disciplines beyond games such as music with Jukedeck's AI-composed music. For a small fee, you can request Jukedeck's AI to compose royalty-free music for your video or vlog. Located in London, UK, Jukedeck continues to drive its research and development by staffing their company with not only computer scientists but with musicians and musicologists to train their artificial intelligence. Their model starts with the customer request for a song with a user-specified mood, duration, types of instruments, and tempo. Jukedeck's AI will compose the music and, with sophisticated music synthesis, generate the audio track. Users make their requests through Jukedeck's online interface, and after about 30 seconds, a track is delivered as an MP3. The user can request another piece if they do not like the first one. At the time of writing this book, the pricing was quite reasonable for royalty free music—$7 a track for individual users and $15 per track for larger businesses. If you like the piece, you can pay $150 for exclusive, permanent rights to it. Jukedeck has not won a Grammy Award, but according to the CEO, Ed Rex, Jukedeck offers an affordable way to add music content. Other systems have been developed to generate music based on established forms such as classical or jazz. David Cope, who is among other things a composer, author, scientist, and AI researcher, developed a system called Experiments in Musical Intelligence, or EMI, that composes classical music. He has commercially recorded some of the EMI-generated music, and he has co-composed music with another AI system he developed called Emily Howell.

Beyond music, the visual arts remain a high-water mark for human creativity and expression. Not only do famous paintings such as Leonardo Da Vinci's *Mona Lisa*, Monet's *Water Lilies*, or Andy Warhol's *Campbell's Soup Cans* immediately come to mind as creative masterpieces but also the movements in art that they represent such as the Renaissance, Impressionism, and Modernism respectively. The creativity and expression in a masterpiece and the exploration that drives new movements in art and culture today seem far from the grasp of even the best in artificial intelligence. By the same token, AI researchers have strived for many years to develop machine-generated art. An early pioneer in the field, Harold Cohen, built a system called AARON. In development since the early 1970s, AARON produces images of people, plants, and other objects. Some of the pictures look interesting but arguably a little bit like paint-by-numbers pieces. *Algorithmic art*, or computer art, has been evolving and has several different classes such as *fractal art*, which takes a shape and uses mathematical formulas that repeat and expand the patterns. Another type of algorithmic art is called *genetic art*, which represents art that starts with a shape, and the computer repeatedly adds to and changes the shape; think of how a snail shell twists and grows larger, forming an increasing cone. Other attempts at computer creativity such as Google's DeepDream have emerged in recent years. DeepDream uses sophisticated artificial intelligence to generate pictures that the computer composes based on images fed into it by computer scientists or the public through the DeepDream Generator website.[54] According to their website, DeepDream was initially built to help engineers to look inside the *mind* of the computer to better understand how it was *seeing* images it was processing, and what came out were very psychedelic images. In the summer of 2016, at an event hosted by the Grey Area Arts Society, over $80,000 were raised from the sale of DeepDream's AI generated images. Despite the price paid, the works remain the product of a program. Whereas, art emanates from the expression of an artist. DeepDream did not spontaneously get struck with the inspiration to create colorful, psychedelic images, but Monet was inspired to create his famous *Water Lilies*—not programmed to.

Though not at the level of Monet or da Vinci, with *computational creativity*, one can still consider how to benefit from artificial intelligence-driven computer creativity. Those with a lower estimation of creativity like Bernd Schmitt simply concluded in an interview, "What impact do you see the singularity having on your areas of specialty—marketing, branding, and creativity? It's entirely possible that supercomputers may do marketing, branding, and creative tasks."[55] However, other scholars, researchers, and philosophers do not share Schmitt's opinion and consider creativity to be an attribute of human consciousness and that machines will serve most effectively as a powerful assistant or

[54] https://deepdreamgenerator.com/
[55] Frieda Klotz, "Are You Ready for Robot Colleagues?" an interview with Bernd Schmitt, *MIT Sloan Management Review*, July 6, 2016, accessed online July 27, 2017

tool for creative people.

Benefits and Profits from Robots

Work forms a fundamental human need, activity, and requirement. We work to earn money to support ourselves and our families, which confers a true dignity and honor to our labor. The ability to work also provides a sense of purpose that extends beyond the individual and contributes to the health and vitality of a community and a nation. The subject of automation and jeopardizing jobs must be looked at very carefully in light of people as well as businesses and will be dealt with in more detail later in the book. The introduction of mechanization to manufacturing and agriculture has changed how people work in significant ways. Manufacturers have benefited from automation for centuries by increasing production to increase efficiency and competitiveness. Machines in some cases have alleviated human health risks doing jobs such as painting and welding that produce toxic fumes dangerous to people. With the advancement of artificial intelligence, robotics in manufacturing will become more sophisticated, safer, and less expensive than ever before.

The robotic sector continues to grow at an increasing pace. A report from Merrill Lynch titled, "Robot Revolution—Global Robot & AI Primer" estimates that the robotics market will be an $83 billion market by 2020.[56] However, the field of robotics did not begin recently. General Motors deployed the first

---

[56] "Robot Revolution – Global Robot & AI Primer," Merrill Lynch, December 16, 2015, accessed online June 4, 2017

industrial robots in the early 1960s. In 1961, GM added the Unimate electro-hydraulic robot to its assembly line in the Ternstedt Division in Trenton, New Jersey made by Unimation Inc. The Unimate robot that unloaded die-cast auto pieces and welded them on auto bodies was a programmable 4000-pound arm developed by George C. Devol and Joseph F. Engelberger in the 1950s. The pioneering work of Devol and Engelberger spawned a revolution in manufacturing. Moreover, it was not just American manufacturing that began to adopt robotics. Looking to open foreign markets to their robot products, Devol traveled to Asia. As a result of his tour of Japan, Unimation, Inc. would strike a deal with Kawasaki Heavy Industries that would supply thousands of robots and create the foundation for Japan to become a world leader in robotics. Even back in the 1960s, during Japan's rise to become an economic superpower, second only to the United States, Japanese leadership was aware of their aging population, foreshadowing a labor shortage in the coming decades. In fact, combined with a low birth rate and restrictive immigration policies, Japan has embraced the inclusion of robotic assistance beyond the factory.

Robots perform at their best in more controlled environments because moving and navigating uneven or unpredictable territory remains a skill which people still excel at much better than machines. However, if you have predictable environments that robotics can serve, there are opportunities to be had. In that light, because of the flat, predictable nature of glass, window washing lends itself to robotics. Some commercially available robotic window washers by Haier, Ecovacs, and others even retail for home use.[57] The Ecovacs Winbot 850 robotic window washer for domestic use sells for around $350 on Amazon. The robotic window washer can scan the window for size and calculate the fastest, best, and most efficient path to cleaning any framed or frameless window. Window cleaning can be dangerous work, especially for higher stories. The Centers for Disease Control and Prevention, a US governmental agency concerned with human safety indicated that in 2011 in the United States that work-related ladder falls accounted for 113 fatalities and 15,460 injuries.[58] Ladders pose a demonstrable risk to human safety. A window cleaning robot may apply to businesses or homes because it saves the risk of injury or worse from a fall when cleaning domestic or commercial windows.

Robotics continue to increase in complexity while also falling in price. Universal Robots based in Odense, Denmark produces a range of industrial robotics with some being in an affordable enough range for small businesses that have repetitive tasks such as gluing, polishing, and product testing that they would like to automate. Because the Universal Robots robots can sense their environment, the robot will stop operating if it bumps into a person. The ability to detect its environment makes a Universal Robots robot much safer than one that will keep moving even if a person or thing is in the way. Such sensitive robots that do not require a cage are called *cobots* because they work

---

[57] Frank O'Connell, "Inside the Winbot 730, a Robotic Window Cleaner," *New York Times*, November 20, 2013, accessed online October 26, 2017

[58] Occupational Ladder Fall Injuries — United States, 2011, Centers for Disease Control and Prevention, April 25, 2014 / 63(16);341-346, accessed online on March 30, 2018

collaboratively alongside humans. Universal Robots cobots start at $23,000.[59] For businesses that deal with dangerous situations such as mining or rescue or need to access narrow or difficult conditions, Transcend Robotics, a Washington DC-based robotics firm, offers the ARTI3 articulated robot. ARTI3 stands out among robots because it can do what most cannot. It can navigate irregular terrain. ARTI3 has three segments and a pair of treads on each section that enable it to climb stairs and other complex terrains. Starting at $9,995, equipped with a 360-degree camera and remote communications capabilities, the ARTI3 would enhance the range of businesses that need to offer inspections in hazardous environments, making a firm more competitive and safer for its employees. Bob House wrote an article for *INC.* in which he stated, "For small business owners, the benefits of automation go beyond operational efficiency. Going forward, companies that are prepared for an automated future will command higher prices in the business-for-sale marketplace."[60] The business-for-sale marketplace represents business owners buying other businesses or selling their own, and firms with automation will command a higher price.

In agriculture, changing labor forces and immigration issues have resulted in shortages of people willing to work on farms, resulting at times in difficulty harvesting crops from the fields and orchards. In a report for *AgFunder News*, Emma Cosgrove wrote, "A 2012 NRDC [Natural Resources Defense Council] report

---

[59] Entrpreneur, https://www.entrepreneur.com/slideshow/274355#3 accessed online March 10, 2018

[60] Bob House, "How the Rise of Robots Will Affect Small Businesses," *INC.*, June 22,2017, accessed online March 10, 2018

estimates that 20% of produce grown in the US doesn't leave the farm either because farmers can't find enough labor, or because the cost of labor isn't covered by the potential revenue of the crop."[61] Farms have been at the forefront of mechanization and automation for many years, and now the next generation of automation is developing to address the shortage of farm labor. Robotic harvesters continue to develop. In "Agriculture Robots: Four Global Trends to Watch," Aseem Prakash notes that robotic strawberry and apple harvesters continue to improve and can shoulder some of the burdens on the farm. Additionally, he notes that the population of farmers continues to age out of the profession and "most young people don't consider farming an attractive profession, and immigration policies around the world are making it difficult to obtain migrant workers."[62] Automation may fill specific roles on the farm, but the decreasing number of people willing to participate in agriculture in contrast to the growing world population and the accompanying need for more food will put pressure on AI and robotics companies to fill the gap. Such a gap will also open the door to opportunities for innovation and profit in the essential agriculture sector.

Robots give artificial intelligence a body—a physical presence in the world. Robotic assistance continues to transform manufacturing and agriculture. Consider the possibilities of how a trainable robot running on artificial intelligence could change your

---

[61] Emma Cosgrove, "More than Robotics is Needed to Solve Farm Labor Shortage," AgFunder News, July 13,2017, accessed online March 30, 2018

[62] Aseem Prakash, "Agriculture Robots: Four Global Trends to Watch," *Robotics Business Review*, August 11, 2017, accessed online March 30, 2018

business or offer an advantage to a company in which you are working or investing. Success in adopting any new technology hinges not only on the technology but identifying the problem that needs to be solved. In other words, if your business problem is with production efficiency or worker safety, a careful analysis of that issue may indicate that automating that step lends itself to artificial intelligence with robotics. Such an advantage would boost the value of your business and provide a healthier workforce.

Chapter 6
Benefits of AI to Society

The saying, *A rising tide lifts all boats*, implies that economic benefits can be beneficial to everyone. The concept extends beyond economics to society as a whole. The discovery of antibiotics by Alexander Flemming in 1928 and subsequent discoveries of other antibiotics changed the world by offering a cure for certain diseases caused by bacteria such as pneumonia, meningitis, syphilis, and many more. The discovery of antibiotics saved millions of lives and significantly extended human life expectancy like nothing invented before. With the rising use of artificial intelligence in many sectors of business and government, opportunities for all of society to feel the benefits of artificial intelligence continue to emerge. Thoughtful application of AI will require good leadership to choose the applications that provide the greatest benefit to society as a whole. There are many sectors to explore, and we will look at caring for the sick and elderly, healthcare, the law, and more.

At a more tactile and emotional level, since 2003, Intelligent Systems Co., Ltd. of Japan has been manufacturing and selling an advanced interactive robotic baby harp seal called PARO. PARO responds to physical, verbal, light, posture, and temperature stimuli with sounds, movements, and facial expressions to communicate if it is content, happy, sad, or uncomfortable. The robot will let you know if it is happy with sounds of contentment and happy facial expressions. PARO will also make unhappy noises when pet too hard or if it is cold. The FDA approved PARO as a class 2 medical device in 2009. Although initially designed as a companion for the elderly in Japan, Europe, and the United States, PARO serves more often as a therapeutic tool in dementia care. In 2015, a quasi-experimental study demonstrated the effectiveness of PARO in dementia care and facilitation of daily care published in the *Journal of the American Medical Directors Association*. Anecdotal reports in the news describe PARO comforting elderly dementia patients and serving as a useful tool before applying medication. Of course, the ethics of supplementing or even replacing human care with AI and robots need a more in-depth investigation and will be addressed later in the book.

Caring for our elderly remains an essential honor and responsibility for our society. For centuries, families have taken care of their elderly mostly in the family home. However, over the past hundred years in the United States, elder care has increasingly been taken up by public and private institutions. In a paper published in *Health Services Research* in 2002 titled, "The 2030 Problem: Caring for Aging Baby Boomers", the authors, James R. Knickman and Emily K. Snell, project a

135% increase in the population in the United States over the age of 65 in the next two decades.[1] In more tangible numbers, with the baby boomers now entering retirement age at a rate of 10,000 per day, the focus on elder care grows at an increasing rate. Depending on your perspective, the expanding population of elderly in the US and other countries such as Japan and the UK offers unique opportunities for both caution and profit from AI and robotics.

Japan, for example, has known for many years that they will have a growing elder population due to the country's low birth rate. Japan continues to evidence some of the lowest birthrates in the world with just over eight births per 1,000 people according to the World Bank DataBank.[2] In other words, the Total Fertility Rate (TFR) of 1.43 children per woman over her lifetime in Japan puts Japan well under the replacement TFR rate of 2.1. Additionally, Japan has one of the world's lowest immigration rates with immigrants making up 1.7% of their population. For comparison, according to data from the United Nations, the United States population is 19% immigrants and stands as the country with the largest number of immigrants in the world. Add to the low birthrate in Japan the fact that they also have one of the longest-lived populations on the planet and the looming crisis of who will take care of the elderly appears quite serious. When there are fewer young people to provide physical as well as economic care for a country, who can step in?

---

[1] James R. Knickman and Emily K. Snell, "The 2030 Problem: Caring for Aging Baby Boomers," Health Services Research 37(4): 849–884, August 2002

[2] Fertility rate, total (births per woman), World Bank DataBank, https://data.worldbank.org/indicator/SP.DYN.TFRT.IN/

With so many people living long past the age of retirement, the Japanese government in conjunction with academic research and private industry plan to fill the care gap with robots. The Japanese government invests millions of dollars in developing robotic elder care and nursing robots, or *care-bots*, intended to make up for the predicted 500,000 to 1,000,000 million eldercare worker shortfall over the next 30 years. Analysts estimate the global care-bot market to reach $17 billion by the year 2020.[3] Researchers plan to have care-bots take over the difficult, repetitive, or dangerous roles that human healthcare workers do now such as lifting patients from a bed or a wheelchair, assisting with housecleaning, preparing food, and helping with the toilette. Researchers at Japan's RIKEN research facility in collaboration with Sumitomo Riko Company Limited developed a nursing care-bot called Robear. Robear looks like a friendly but somber white bear and weighs 300 lbs. Robear was designed to lift and carry people from beads to chairs or to help patients stand or get into a sitting position. Advanced sensors and mechanics allow such a powerful robot the delicacy of touch combined with the strength to carry adults. Many engineering innovations in Robear will enable it to be more agile and to take up less space than previous nursing care-bots. Effective home care-bots will need to be nimble and thin to maneuver in apartments and rooms at care facilities. The growth of the elder population and growth in robotics is poised to expand significantly over the next thirty years.

At a more personal level, dramatic improvements

---

[3] "Robot Revolution – Global Robot & AI Primer," Merrill Lynch, December 16, 2015, accessed online June 4, 2017

in AI in healthcare continue to change our world from digital monitoring, individualized health care, and enhanced AI diagnostic capabilities. IBM's Watson supercomputer system gained global recognition in February 2011 as the first computer champion of the question answering game show *Jeopardy!*. Watson is a question-answering computer system designed to hear and respond to questions in natural language and is one of the most sophisticated computing systems ever built in the world. It stands as the flagship effort-leading AI at IBM. Watson began with no initial intelligence like a newborn but started to learn and make its judgments about what it was learning. Watson is able to read 2 million books in a minute, which means he can go through the Library of Congress in 18 minutes. Watson processed that information and learned by interacting with humans to understand what is a right or wrong answer. On top of that, it took Watson five years to learn to listen and speak so that it could verbally interact with people. Although winning *Jeopardy!* introduced Watson to the world, it is only the beginning of what its architects envisioned for it. Watson is also involved in healthcare, military, and other projects.

After Watson triumphed at *Jeopardy!*, the researchers at IBM began to look at other applications and have partnered with a number of cancer research centers to use Watson's vast ability to absorb, understand, and make judgments on new medical information. Moreover, Watson recommends treatment options for cancer patients. The project is called Watson Oncology. Watson Oncology began with the intention to help clinical doctors find the very best treatment options for their cancer patients, and Watson brings a unique skill set to oncologists. After a cancer diagnosis, the treatment options often involve a whole team of experts to understand the type of cancer, the genetic data from the patient, and the types of treatments available either as standard care or more cutting-edge treatments still being studied in clinical trials all around the world. As the clinical oncologists and scientists must spend time caring for patients, they must also try to keep up with the all the latest literature, conferences, and trials going around the world.

In training Watson to be a *doctor*, the supercomputer learned to read medical literature and read the over 25,000,000 medical journal articles plus books and clinical trial information. To acquire this base of research, it would take one person reading one article every 30 minutes working 12 hours a day a total of 2,854 years to learn all that information. However, medical literature is growing every day. PubMed, the collection of biomedical literature maintained by the National Instituts of Health contains over 26 million records and adds another 500,000 records a year or 1,300 per day! No researcher can keep up with this amount of new literature, but Watson can handle it in seconds. With its incredible knowledge base, Watson went to work with the clinical experts and would learn treatments and recommend them to the doctors based on the latest literature and the patient's clinical data. Watson also reviewed, using machine learning, the patients MRI and CT scans to help the team decide the best course of action.

A clinical study compared the cancer treatment recommendations made by Watson without any human assistance with those made by experts in the field. Watson scored 95% accurate which alone is impressive, but what shocked the researchers was that 30 percent of

Watson's suggestions were new treatments not thought of by the experts that offered hope for alternatives where treatment options had run out. According to the Sloan Kettering Cancer Research Center, 90% of lung cancer treatment recommendations are now coming from Watson! The immense power of AI assistance for cancer treatment teams only begins to suggest the potential of Watson and undoubtedly other solutions for harnessing the rapidly expanding world of knowledge that no one person could ever be able to absorb and use in one or many lifetimes.

Opportunities in medicine to enhance doctors' ability to diagnose and treat patients may eventually drive down the cost of healthcare. Moreover, as we learned earlier, certain types of surgery use robotics and artificial intelligence to be less invasive. Robotic surgery brings the advantage of smaller incisions, more precise manipulations, and 3D visualizations that put eyes where traditionally a surgeon could not see. Using remote surgery, doctors can extend their ability to help patients literally around the world. The application also could have a profound effect on the battlefield. In an article by Gary Martinic titled, "Glimpses of Future Battlefield Medicine—the Proliferation of Robotic Surgeons and Unmanned Vehicles and Technologies," the author notes that nearly 90% of battlefield deaths occur in the first 30 minutes following the injury.[4] Such a statistic underscores the urgency of medical attention and the potential advantage of using battlefield robotic

---

[4] Gary Martinic, "Glimpses of Future Battlefield Medicine – the Proliferation of Robotic Surgeons and Unmanned Vehicles and Technologies," *Journal of Military and Veterans' Health*, Vol. 22 No. 3, 2017

surgeons remotely operated to address life-threatening injuries faster than currently possible today. As mentioned earlier, doctors have successfully performed a variety of remote surgical procedures, including heart surgery, gallbladder removal, and liver operations. Research and development in battlefield surgeons by the US Army and DARPA continue to make progress by developing robots capable of going into the active battle zone to pick up fallen soldiers and carry them away from the field of battle for treatment.

The Law

Our collective perception of law comes mostly from movies, novels, and TV shows depicting tense courtroom dramas with sharp lawyers and judges applying the law to put bad guys in jail or to protect the innocent from bullies, outlaws, and evil corporations. However, the realm of law extends far beyond crime and punishment. Laws guide, protect, and shape our culture's attitudes. Laws dictating environmental regulations, minimum wage, occupational safety, drug development, air traffic, and so on touch every aspect of our lives. Just as in every other facet of life, AI, machine learning, and robotics will change and in some cases, transform law. Moreover, law will not only be affected by these new technologies in many different ways, but these technologies will be shaped and governed by law. Long-standing laws regarding privacy, liability, and freedom will face challenges. Understanding the evolving relationship between law and technology will influence your strategy towards profiting and protecting yourself from AI. In this chapter, we will get a sense of how AI and robotic technologies are changing the law in three major areas:

◊   The practice of law

◊   Law adapting to machine learning and robotics

◊   Robo-attorneys

Some of the earliest written law, circa 1930 BCE, are clay tablets written in Sumerian and ascribed to the wise shepherd, farmer, and later king, Lipit-Ishtar, in what is today modern Iraq. The prologue of the Sumerian clay tablets, The Code of Lipit-Ishtar, suggests that Lipit-Ishtar was called by the gods Enlil and Ninlil to write these rules and punishments to bring peace to the lands of Sumer and Akkad. The Code of Lipit-Ishtar contained strong penalties and fines for the various crimes described. The Code even includes the earliest known law on child support. It reads as follows:

"If a man's wife has not borne him children, but a harlot from the public square has borne him children, he shall provide grain, oil, and clothing for that harlot. The children which the harlot has borne him shall be his heirs, and as long as his wife lives the harlot shall not live in the house with the wife."[5]

The clay tablet fragments containing the Code of Lipit-Ishtar, although broken and incomplete, contained at least 38 laws. In the subsequent 4,000 years, man has created more sophisticated legal systems and hundreds of thousands of laws governing all aspects human behavior. Remarkably, no one knows how many laws there are today. In 2011, *The Wall Street Journal*

---

[5] James R. Court, *Codex Collections from Mesopotamia and Asia Minor*, Scholars Press, 1995.

published an article detailing the plight facing legal scholars for decades—how many criminal laws are in the United States Federal Code?[6] Although the Federal Government has over 23,000 pages of law, some laws in these pages are amended, and others were repealed, so it is not just a simple case of counting the laws. Back in the 1980s, the Justice Department tried to count all the laws and after two years finally gave up. Before they finally abandoned the project, they offered a guess that there are about 3,000 criminal laws on the books in the US, but this was only approximate. Now, remember that this effort only looked at the criminal code and did not even attempt to look at all the federal regulations. Moreover, Congress gives the executive branch of the United States government the right to make additional rules to support and enforce the many laws passed by Congress. The process is called *rulemaking*. Federal regulations carry criminal penalties, and the estimated number of federal rules easily reaches up to the tens of thousands. Law has become massive and complex, and piling on top of that all the court rulings, opinions, and legal scholarship, no one person or even team can keep up with all of the growing and evolving information. In the face of such burgeoning information, it makes sense that AI is stepping into the breach.

In collaboration with IBM Watson, a company called ROSS Intelligence offers what the headlines splashily called, *the first artificially intelligent attorney*. ROSS started out as a research project at the University of Toronto to enhance attorney effectiveness with artificial intelligence by providing better, faster

---

[6] Gary Fields and John R. Emshwiller, "Many Failed Efforts to Count Nation's Federal Criminal Laws," *Wall Street Journal*, July 23, 2011

legal research. In 2015, ROSS moved to Palo Alto, California where it began its 10-month training in bankruptcy law. ROSS then entered the marketplace, and the Ohio law firm Baker Hostetler acquired the first license for ROSS. One might think of Watson as some massive supercomputer housed in a facility full of blinking lights and wires, but Watson comprises a growing number of applications in the cloud. ROSS uses the artificial intelligence capabilities of Watson's question and answer component just like in the famous victory on *Jeopardy!*. ROSS uses three capabilities in Watson to derive an answer to questions. First, Watson determines what the question is asking. Once the question becomes clear to Watson, it retrieves relevant documents from its legal database. And finally, Watson, using a variety of AI and statistical techniques, ranks its answers to find the best one. Watson is famous for devouring information in the order of millions of documents in days, but the last part where it must rank the answers comes from expert humans training the system, building training sets to teach Watson. The teaching takes a long time. In the same vein, it took ten months of human training to train ROSS Intelligence in bankruptcy laws. After such a huge effort, ROSS allows attorneys to ask ROSS questions by speaking, and ROSS will return a relevant answer in English. The power of ROSS comes from its ability to return appropriate responses to legal issues in the bankruptcy space and to continue to learn by consuming legal opinions, decisions, and changes in the law saving thousands of hours in document review.

Document review represents the process for discovery in government investigations or for litigation. Attorneys review, in some cases, thousands and even millions of documents looking for information relevant to their case or relevant to their opponent's requests for information. Certain types of cases require large amounts of documents such as mergers and acquisitions, litigation, and audits. The document review may have several classes of information; the attorney will tag the documents with terms such as relevance to the case at hand or whether the information is confidential or otherwise privileged. The "Litigation Cost Survey of Major Companies[7]" presented at the 2010 Conference on Civil Litigation at Duke Law School presented staggering numbers. According to the report in 2008, a major lawsuit going to trial in the United States presented an average of 4,980,441 pages of evidence. The cost of litigation increased by 78% from 2000-2008, not due to increased fees but the higher burden of discovery. Large companies report paying discovery costs ranging from a half a million to nearly ten million dollars for every major trial. Such a time and financial burden undoubtedly invite automation innovation with a system like ROSS.

An increase in automated, online dispute resolution through companies like Modria based in San Jose, California may have a profound effect on lawyers and judges if more disputes go through these services, cutting out traditional avenues. Colin Rule founded Modria in 2011. Before starting Modria, Colin led the development of eBay's automated dispute resolution system that resolves over 60 million disputes a year for eBay clients. The system at eBay and others like it guides disputants through a series of questions

---

[7] "Litigation Cost Survey of Major Companies," 2010 Conference on Civil Litigation at Duke Law School, 2010, accessed on line 2/19/2017

and, based on calculations, proposes resolutions. If no decision arises, human adjudicators come into play. In the United Kingdom, a student developed an application that helps people dispute parking tickets. The chatbot called DoNotPay is a free application developed by Joshua Browder. The system uses artificial intelligence to guide a dispute resolution algorithm that has helped over 160,000 people reverse a parking ticket, resulting in the recuperation of over $4 million. Proponents of the free service laud the application of AI to help resolve parking tickets without having to take time off to go to court or hire an expensive lawyer. Outside of the United States, other countries are developing similar systems for a more complex automated dispute resolution such as a system in the Netherlands called Rechtwijzer that helps couples separate or divorce without the intervention of the court.[8] According to the article, "Rechtwijzer Launches in the Netherlands," "Partners login to the platform from their own computers, describe their situation and make proposals for agreements. It is only possible to proceed to the next step if both partners agree with each other. If the partners cannot agree on any one term, there is an option to use a mediator or arbitrator."

Advocates claim that automated dispute resolution systems increase our access to justice. According to an article in the *America Bar Association Journal*, a study of the efforts of automation in three courts in Michigan, "and 17,000 cases revealed a 74 percent reduction in average days to case resolution."[9] It will take more

---

[8] Rechtwijzer Launches in the Netherlands, *HiiL Inovating Justice*, November 26, 2015, accessed online October 26, 2017
[9] Anna Stolley Persky, "Michigan Program Allows People to Resolve

time and analysis to know if online dispute resolution truly profits the general public. Ayalet Sela writes in the *Cornell Journal of Law and Public Policy* that, "Proponents of ODR [online dispute resolution] argue that it is a natural next evolutionary step, suggesting that technology can make many processes more accessible, less expensive, easier, and faster to complete; and that it entails new features that can improve procedural quality. Critics of ODR contend that judicial and ADR [automated dispute resolution] processes cannot be adequately conducted online and that the claimed efficiencies of ODR come at the expense of procedural quality, primarily due to the limitations the online environment imposes on human communication, privacy, confidentiality and neutrality."[10]

Expanding our Presence

Another area of robotics poised to grow is telepresence robots. *Telepresence* refers to the combination of robotics, virtual reality, and telecommunications that projects your presence in a location distant from your own. In some respects, having a phone call, video conference, or even an email projects your presence over distance, but phone calls, video conferencing, and emails lack a physical embodiment. The phone or video screen remains stationary, but a telepresence can move

---

Legal Issues Online," *ABA Journal. American*

Bar Association, 1 Dec. 2016, accessed online March 10, 2018
[10] Ayalet Sela, "Streamlining Justice: How Online Courts Can Resolve the Challenges of Pro Se Litigation," *Cornell Journal of Law and Public Policy*, Vol. 26:331, 2016 P.334-335

around the office or home. Telepresence bots have been described as an iPad on a stick with wheels. Companies such as Giraffe, VGo, and Double Robotics make telepresence bots that have a screen on the front that displays a remote user's face. The bot is equipped with a microphone, camera, and speakers so that the distant person can talk and interact as if they were in the same room but can be thousands of miles away. Again, this is not entirely different from FaceTime or Skype, but it has one big difference—the screen sits atop a robot body controlled by the remote user. The display can roll almost silently through the office or anywhere with a smooth surface.

Manufacturers of telepresence robots such as Double Robotics, which sells the Double 2 Telepresence Robot for about $2500, claim that their robots provide the customer with a physical presence at work or school when he or she cannot be there in person. Some of the applications of telepresence bots sound very useful. Double Robotics provides a case study on their website of a student battling cancer that uses a telepresence robot to attend high school, interacting with teachers and students in real time, even *walking* the halls between classes and talking with friends. The case study remarks that the telepresence engages the class, the user, and the instructor more effectively than a simple online course. Remote education adds a positive dimension to the use of telepresence. Other positive telepresence uses include reduction of travel budget for companies that need collaboration among teams working at different locations. Additionally, some doctors conduct patient visits halfway around the world looking at charts and communicating with patients in real time, expanding a practice to patients that would otherwise be impractical or even impossible.

Apart from the positive aspects of telepresence mentioned above, some articles and videos deal with the social rules for dealing with a telepresence bot. Imagine you are working at your desk and instead of hearing your boss's footsteps, you hear the soft whine of servo motors whirring. The telepresence bot pauses behind you. You look over your shoulder to see your boss's face on a screen peering at you and asking if you finished the report she wants. A video by *Wired* deals somewhat humorously with establishing standards for people coping with telepresence bots and for the remote bot drivers coping with office workers.[11] With telepresence bots, that scenario can happen anytime. One of the rules suggested states that you should not touch a telepresence bot or move it without first checking with the person driving the bot remotely. It just seems so tempting to just stash the bot when you are tired of it wheeling around the office, but we are reminded that you should treat the bot as if the real person were there. In other words, follow the same code of conduct you would if it were a real person.

Telepresence robots have entered the office, hospitals, and schools as a way for a person to exert a physical presence from a remote location. Students can attend class and *hang out* with their friends when circumstances prevent their ability to attend school. Many other applications of telepresence bots extend a physical presence over distance, but with this new technology new rules and norms will be needed in the office, hallways, and playgrounds for dealing with

---

[11] https://youtu.be/ho1RDiZ5Xew

the telepresence. In many ways, the evolving code of conduct for telepresence bots is paving the way for human-robot hybrids. Combining the robotic advances seen in Robear with telepresence may provide a deeper sense of connection with an elder member of the family in a nursing home that is currently not possible. It may sound too futuristic, but imagine *helping* your parent to cook dinner and get ready for bed remotely through the strong arms of a robot with telepresence.

As artificial intelligence grows, so does the field of robotics. Some of this market will be for industrial and military applications. Robotics will also provide an opportunity for enhancing our lives in many new ways. Some are in the areas of convenience such a robotic domestic help. Already on the market for some time, Roomba the indoor vacuum, works autonomously to vacuum floors, and it will even go back to its charging station when it is low on energy. People with disabilities will see more opportunities to expand their world with robotic assistance. Domestic help remains a hot area of research and development to make intelligent robots that can help with any number of domestic chores from food preparation, medication reminders, bathing, and mobility assistance. According to an article on Robohub titled, "Robots Pick up the Challenge of Home Care Needs" a growing sector in domestic robots is companion care. *Companion robots* are emerging as a specific category. They are designed to act as interfaces that will make it easier for the elderly to enrich their social lives and connect with their families and friends."[12] One such robot available

---

[12] Peter Feuilherade, "Robots pick up the challenge of home care needs," Robohub news, April 12, 2017, accessed online October 30, 2017

in Japan called Kuri is a small robot on wheels with a cone shaped body and a smooth round head with circular eyes that respond to touch, sound, and visual cues and is designed to keep people company. Kuri communicates with a robot language of *beeps* and *boops* that people can quickly learn. Moreover, Kuri captures pictures and movies to save fun memories.[13] Retailing for about $700, Kuri can brighten up life at home. Robots also will help disabled people to walk. With a new exoskeleton, robotics amplify one's leg strength or can assist a person with robotic exoskeleton arms and back to help with yard work or around the house.

Robotics and AI together stand poised to alter many sectors of the economy and culture. No discussion of their impact would be complete without addressing autonomous vehicles. Numerous articles and news stories feature self-driving cars. Taxi and Uber drivers worry that autonomous vehicles will permanently take away their livelihoods. People express concern that the auto-piloted cars will not be safe, or people may lose the freedom of the road that so many value and see as a right. Given the many concerns about the culture-changing effects of self-driving cars, the autonomous passenger vehicle has already arrived, and the technology behind these vehicles is remarkable. Artificial intelligence, advanced sensor technology, and robotics combine to drive passengers through complex and changing situations. Whether you look forward to or dread the driverless car, it is here, and in some ways, our driving culture will change. Large automotive companies already have been investigating a future

---

[13] Kuri, Company website, Accessed online October 30, 2017

where individuals or families no longer purchase their car.

Originally founded in 2000 in Boston, MA, Zipcar offers on-demand car rentals accessible through a smartphone app, and the cars wait for renters around cities in designated Zipcar parking spaces. Zipcar has altered drivers' relationships to car ownership by offering the convenience of car ownership without the responsibility of maintaining or insuring a private car or having to go to a car rental office. Zipcars can be found accessibly located in many major cities, and the user may reserve a car for a specific time for a monthly or annual fee. Zipcar claims in an infographic that one Zipcar replaces thirteen personally owned cars.[14] With over 1,000,000 subscribers worldwide and over 10,000 ZipCars on the road, their car sharing business has removed hundreds of thousands of vehicles from crowded cities.

Zipcar makes an excellent example of something known as the *sharing economy*, and the ridesharing services such as Uber and Lyft provide the best example of the sharing economy. Uber and Lyft drivers own and maintain their vehicles and by contract cannot be hailed from the street like a taxi. Uber drivers are sharing their vehicles and time to transport people around town. Uber picks up a ride request by smartphone, and using GPS, it determines the nearest ride options, offers the ride requestor the details of the driver, and the expected wait time. No money changes hands as it is all handled by Uber electronically through smartphones and electronic payments. Uber's app, powered by machine

---

[14] ZipCar infographic accessed on July, 2, 2017http://www.zipcar.com/ziptopia/inside-zipcar/zipcar-hearts-the-earth

learning, calculates the nearest driver and hales them. Uber has transformed the sharing economy in unique and unexpected ways. A recent study by researchers at Oxford University, in the UK including Carl Benedikt Frey cast a unique light on the effects of Uber on jobs and wages in several American cities.[15]

The study titled, "Drivers of Disruption? Estimating the Uber Effect," to the surprise of the press and of the taxi drivers who have protested Uber and other ridesharing businesses, demonstrated that the ridesharing service did not destroy the taxi business or other public transportation jobs. In fact, the authors found that Uber upon introduction to a city increased the number of self-employed people by 50% and increased the number of taxi drivers. The study did note a 10% decrease in earnings for salaried taxi drivers, but a 10% increase in earnings for self-employed taxi drivers offset the taxi driver losses. The net benefit of Uber as a sharing economy platform appears to be positive in creating more employment and not bringing the prophesied destruction of the taxi business in American cities.

Airbnb, the home sharing company that allows people to offer rooms or even their whole homes to guests online through their application, was initially considered a threat to the hotel industry. Interestingly, in an article from *Business Insider* titled, "Airbnb Might Not Be Hurting the Hotel Industry after All," cited a report by the industry analysis group, STR. It looked at Airbnb in contrast to major hotel chains such as Marriot, Hilton, and several others in 13 different

---

[15] Thor Berger, Chinchih Chen, and Carl Benedikt Frey, "Drivers of Disruption? Estimating the Uber Effect," January 23, 2017

cities. Apart from the fact that Airbnb has over 2.3 million rooms listed, which is more than twice the number of rooms listed by any hotel chain, the house sharing application did not negatively impact hotel occupancy.[16] Moreover, the analysis shows that the Airbnb traveler is looking for a different experience than what the typical hotel offers. The uniqueness of staying in a home with its quirks, meeting local people, and location in neighborhoods not occupied by hotels has expanded travel options without actually denting the hotel industry. Moreover, in an article in *The Wall Street Journal*, Rachel Botsman notes that, "Economic impact studies by Airbnb found that more than 10,000 of its hosts have used rental income to support themselves while launching a business." As labor markets shift, Airbnb and more generally the sharing economy offers new, accessible streams of income for providers and alternative experiences for the consumer.[17]

Uber and Zipcar serve not only as examples of a growing sharing economy, but they help to illuminate the destruction of old, traditional ways that we move people from their homes to the airport or from the restaurant to their hotel. With the rise of autonomous vehicles, more will change. Uber along with Google, Ford, Tesla, and others, envision a day that robotic cars replace human drivers altogether. As one looks to profit and protect from these changes, one must take a disciplined approach to thinking about the whole

---

[16] Avery Hartmans, "Airbnb might not be hurting the hotel industry after all," *Business Insider*, October 12, 2016, accessed online October 26, 2017

[17] Rachel Botsman and Andrew Keen, "Can the Sharing Economy Provide Good Jobs?," *Wall Street Journal*, May 10, 2015, accessed online October 27, 2017

picture. If fewer people buy private vehicles, consider the effects on the automotive insurance agencies, automobile financing, automotive repair shops, rural areas less well served by ride sharing, and even the era of automobile design. Will it matter how the autonomous vehicles look? Maybe the real innovations will come with how cars clean themselves after one set of passengers leaves the car and another embarks. Will the auto dealership be relegated to the charming and quaint world of equestrian services, making saddles and bridles for a small remaining set of the population that still can afford the luxury of a human-driven car for fun or competition?

Chapter 7
Artificial Intelligence in Our Personal Lives

Machines are already picking your hookups, dates, and mates. Machine learning continues to permeate all levels of society, government, and commerce. Chapter Seven seeks to provide the tools and insights to see how you can use artificial intelligence and robotics to your advantage. One question to bear in mind that has been repeated in this book is what are the problems facing me? Identifying problems, also called *pain point*s, will help us to explore how AI already influences your personal life as well as how to identify where AI and robotics services may help you in your personal life. In other words, actively think about ways that artificial intelligence will help you amplify your power. Just as the tractor amplified the power of a single farmer to work much larger pieces of land, or the airplane significantly reduced the time to travel across and between continents, think of how you can use artificial intelligence to amplify your power and relieve pain points in your life. In other words, consider the special abilities you have that will help you cope and even thrive in an AI-enabled world.

Dating

The beauty, the agony, the struggle, the chase, and the process of finding a mate defines a central aspect of the human experience. Naturally, such a primary focus of our lives has also been the subject of many of our most favorite stories, novels, books, movies, and plays. These stories that range from comedy to tragedy begin often with how people first meet. The meeting may happen as an arranged marriage, a chance encounter, a blind date, a friend's recommendation, a meeting of opponents with a particular chemistry, or a matchmaker. Bringing two people together often requires a mix of luck and perseverance. Think of Cinderella meeting the prince by enchanted means or the mixture of media formats such as radio and snail mail in *Sleepless in Seattle* or email in *You've Got Mail*. The classic musical *Fiddler on the Roof* employs a matchmaker named Yente playing a central role in setting off the argument as to whether a girl should be able to choose her husband. Regardless of how people meet, there needs to be a spark to make a successful connection. Human matchmakers consider many factors when making a match like temperament, ambitions, beauty, attitude, religion, socioeconomic status, and such to make the best match. According to Carrie Ritchie of the *Indianapolis Star*, human matchmakers remain popular with singles in the United States. However, the rising star in matchmaking is the artificial intelligence that lies behind the scenes on internet dating sites. Online dating ranks as the second most popular way to meet people in the US behind the old-fashioned way through friends and family. The variety of dating sites continues to expand from the well-established eHarmony and Match.com to Tinder and Bumble. Over 40 million Americans have used an online dating site, and the trend is going up across all sectors and age groups with significant increases in both the younger ages from 18-24 (22%) and the older people from 55-64 (5%) according to the Pew Research Center.

The growth of online dating in more senior age groups reflects a positive impact of technology in helping people to meet a fundamental need for intimacy and connection. Intimacy, both emotional and physical, give life meaning and significantly impact our health and wellbeing. Studies show a significant correlation between loneliness and depression in the elderly.[1] Encouraging seniors to try online dating as another way to open their social lives and try to find a partner if single. Studies demonstrate that physical intimacy for people over 70 continues throughout their lives according to a research article published in *The New England Journal of Medicine* titled, "A Study of Sexuality and Health among Older Adults in the United States."[2] In fact, the authors showed that the frequency of sexual activity among older people up to the age of 75 was the same as reported by people in the

---

[1] B. H. Green et al, "Risk factors for depression in elderly people: A prospective study," *Acta Psychiatrica Scandanavica*, 86(3):213–7,1992
[2] Stacy Tessler Lindau et al., "A Study of Sexuality and Health among Older Adults in the United States," *New England Journal of Medicine*, 357(8): 762–774, Aug 23, 2007

18-59 years old population. The many health effects of sex, including lowering the risk of heart attack, better sleep, and lower rates of depression also hold for older people as well according to Roni Beth-Tower, Ph.D. in the article titled "Benefits of Sex After 50."[3] With seniors reported to be the fastest growing segment using online dating, it appears to be successful for finding a sexual relationship for seniors. According to research by Sue Malta, Ph.D., all the seniors in Australia who met online reported being in a sexual relationship.[4]

Clearly, internet dating continues to rise in popularity, but the question remains as to the success of internet dating. If successful means a marriage ultimately, then internet dating has continued to improve. A study conducted at the University of Chicago and published in a paper titled, "Marital Satisfaction and Breakups Differ Across Online and Offline Meeting Venues," concluded that more than one-third of all marriages in the US started online between the years of 2005 and 2012. The study went on to find that the marriages that began online were, in fact, less likely to end in divorce. Online marriages over the length of the study stayed together 94% of the time versus 92.4% of the time for analog matches. Additionally, online marriages reported a higher satisfaction score than marriages from other venues. Now, it must be noted that eHarmony funded this study, but the authors claim to have had full independence in data collection

---

[3] Roni Beth-Tower, "Benefits of Sex After 50," *Psychology Today*, July 28, 2017, accessed online March 12, 2018

[4] Sue Malta, "Intimacy and Older Adults: A Comparison between Online and Offline Romantic Relationships," Refereed paper, Sociology, Faculty of Life and Social Sciences, Swinburne University of Technology, 2018, accessed online March 12, 2018

and analysis. An article in UChicagoNews.com by William Harms speculated as to the cause of the difference in marriage success rates in the University of Chicago study, citing the population online may be more focused on finding a marriage partner than the general public in offline venues. Also, the number of people that you can check out online far exceeds the physical limitations of a gathering, bar, or coffee shop. Apart from the sheer volume of possibilities to virtually meet hundreds of pre-selected potential partners online from the comfort of your own home, another aspect of online dating that the user cannot see is the use of powerful artificial intelligence and machine learning to suggest connections. In essence, AI is a computer matchmaker, a mechanized Yente. Most sites claim to have proprietary computer programs that match users based on answers they gave in questionnaires. After crunching a client's preferences and personality profile, the computer starts the process of finding a match based on similar interests and traits. Additionally, as users select people to contact, the machines learn more about one's preferences. Using your personal information and preferences, the suggestions of who you might date become more refined. The method is called *machine learning*. In some respects, it is the same as what you experience when you see Amazon or Netflix recommend products and movies based on what you and other people who bought what you are buying. Online dating works similarly.

Already, online dating companies are technology companies that use massive computing power to support the millions of singles looking for love across the nation. I cannot help but wonder if the modest improvement in matchmaking seen online over other venues will continue to grow. Machine learning performs better with more data. Perhaps people will be willing to give up more of their data than just personality questionnaires; maybe also sharing one's spending habits or data from a Fitbit will make even better matches. Given the current trend of increasing data, we should anticipate computers to be better matchmakers with an appetite for more and more data.

Far beyond matchmaking, AI already permeates our work, personal lives, and society in many other ways. With Siri on our phones, Alexa in our houses, and Amazon and Netflix online, we without question continue to adapt AI tools and services into our daily lives and often with no perceivable resistance. When does technology become so fully adopted that it slips into the background of our consciousness? In "The Disappearance of Technology: Toward an Ecological Model of Literacy," the authors, Bertram C. Bruce and Maureen P. Hogan, describe the embedding of technology in our lives as a process by which something like the telephone or electricity begins as something new, external.[5] However, over time the technology starts to embed itself in our lives and habits. In other words, the technology fades from our consciousness and become the fabric of our lives. Think of how little thought one gives to electricity or the telephone. Similarly, the smartphone, GPS-enabled navigation, and Amazon and Netflix recommendation-engines are powered by AI. In many respects, AI already has

[5] David Reinking, Michael C. McKenna, Linda D. Labbo, Ronald D. Kieffer, *Handbook of Literacy and Technology: Transformations in A Post-typographic World*, edited by, 1998, Laurence Erlbaum Associates, Mahwah, New Jersey London

become embedded in our lives today.

Personal Assistants for Life Management

The personal assistant often gets depicted in movies and television as the overworked shadow of an important person. That shadow keeps all the little things from falling through the cracks, anticipating needs before the boss even thinks of them. The personal assistant essentially smooths out a hectic life, bringing order, and fulfilling obligations like remembering an anniversary or a birthday. At some point, everyone has wished for that personal assistant for their own lives, and now through artificial intelligence and the wide availability of broadband internet, digital personal assistants are cropping up everywhere (even when you may not want them). They may not be as capable as a real human assistant, but they are easy to find and work quite well. Earlier, we have looked at digital assistants available through smartphones and computers such as Apple's Siri, Microsoft's Cortana, and Google Assistant. However, there are many more digital assistants available with names like Jibo, Bixby, and SILVIA.[6] SILVIA is an artificial intelligence system that holds conversations with people through computers and smartphones while simultaneously displaying a female-featured head that blinks and shows some emotional signs on its face. Many of these assistants will:

◊ Answer questions on the fly

◊ Keep track of calendars

---

[6] "Top 22 Intelligent Personal Assistants or Automated Personal Assistants," Predictive Analytics.com, accessed online October 28, 2017

◊ Make time-based reminders

◊ Make location-based reminders

◊ Provide real-time updates on favorite sporting events, news stories, etc.

The personal assistant has also left the computer or smartphone and now lives among peoples as a continuous presence in the home with Amazon's Echo and Google Home. These assistants, such as Echo, come as a voice-activated smart speaker for home use that responds to the name *Alexa*. Echo contains speakers and microphones that connect to the internet to deploy its voice recognition capabilities to play songs, check traffic or the weather just by calling out Alexa's name. You wake Echo by calling out its name, *Alexa*, and then ask a question. Amazon's artificial intelligence analyzes your question and returns an answer or action. The commands can be more than requesting to play a song or to get the news; Echo can also interact with smart devices in the home like smart locks and even order products or food from a restaurant. Recently, in the highly competitive field of digital personal assistants, fierce rivals, Microsoft and Amazon, have announced a collaboration that will allow Microsoft's personal assistant Cortana to talk to Amazon's Echo.[7] Echo has focused mainly on assisting in home-related tasks while Cortana developed strengths in email, calendar, reminders, and such. Now the companies will cooperate and allow someone to ask Alexa to call Cortana or Cortana to call Alexa, effectively taking advantage of

---

[7] Jay Greene and Laura Stevens, "Amazon's Alexa and Microsoft's Cortana Will Soon Be Able to Talk to Each Other," *Features, Dow Jones Newswires*, August 30, 2017, accessed online October 29, 2017

each system's strengths. From a smartphone or a laptop computer, someone can call Alexa and ask to turn the heat up at home and order a pizza. Digital assistants are becoming more capable and now even cooperative with other digital assistants.

Coping with Artificial Intelligence

Preparing yourself to take advantage of and to cope with the expanding effects of artificial intelligence in your life involves understanding your humanness and what makes people unique from artificial intelligence. The fact that you are reading a book on how to profit and protect yourself from artificial intelligence already shows your interest in one of the most disruptive technologies of our times. Knowing about AI and its many applications helps you to focus on developing yourself to thrive in the new world, and as just importantly, how to manage the constantly changing landscape. Machines may perform certain tasks faster and more effectively than people, but certain human qualities such as empathy, creativity, and judgement are more valuable than ever. With machines taking over certain tasks, we can focus more on skills such coaching and mentoring. Moreover, with the digital world occupying more of our time and space, taking steps to maintain a healthy sense of self becomes more important than ever.

Artificial intelligence performs many tasks that take in information, analyze it, and produce a prediction. For example, facial recognition works by analyzing a picture and predicting to whom the face belongs. Other examples include autonomous cars, language translation, and scheduling. In an article in the *Harvard Business Review* titled, "How Artificial Intelligence Will Redefine Management," the authors show that 54% of a manager's time spent on the job goes to administration such as juggling employee schedules and writing reports, and these tasks will soon be automated.[8] With more tasks being automated, other responsibilities such as problem solving, developing people, and collaboration will occupy more of a manager's time. The skills to exceed at home and at work depend on emotional qualities like empathy and judgement. Megan Beck and Barry Libert suggest in "The Rise of AI Makes Emotional Intelligence More Important" that cultivating emotional intelligence will bring great value in an AI-driven world. "Those that want to stay relevant in their professions will need to focus on skills and capabilities that artificial intelligence has trouble replicating—understanding, motivating, and interacting with human beings."[9]

Staying on top of how AI-driven systems like social media affect our psychology will help you to better cope with their effects. In 2014, Facebook landed on TV, cable news, and the front page of many newspapers following a publication in the *Proceedings of the National Academy of Science* detailing the results of an experiment conducted on Facebook users. The article titled, "Experimental Evidence of Massive-Scale Emotional Contagion through Social Networks," describes how researchers at Facebook, Cornell

[8] Vegard Kolbjørnsrud, Richard Amico, and Robert J. Thomas, "How Artificial Intelligence Will Redefine Management," *Harvard Business Review*, November 2, 2016, accessed online March 12, 2018

[9] Megan Beck and Barry Libert, "The Rise of AI Makes Emotional Intelligence More Important," *Harvard Business Review*, February 15, 2017, accessed online on March 12, 2018

University, and University of California, San Francisco used computer algorithms to manipulate individual news feeds on Facebook to see if they could affect Facebook users' emotions.[10] Previous sociology research had shown that emotional states like depression or happiness transfer among personal social networks in a process called *emotional contagion*. Although researchers accepted the concept of emotional contagion among people directly interacting in the physical world, they did not know if emotional contagion worked over social media. Facebook, as one of the world's most extensive social networking and service companies with over two billion users worldwide, commanded a unique position to test whether emotional contagion would work through social networks on computers and smartphones. Facebook users post messages, links, photos, and videos that get shared with their friends. Their friends see these posts on what Facebook calls a *news feed*. On the news feed, people express their emotions about their friends' posts through written comments, sharing of content, and *liking*. Facebook uses computer algorithms to decide what gets posted on everyone's news feed. The algorithm tries to put topics that will engage users and fit what people like based on past preferences. In other words, Facebook manipulates your news feed to keep you engaged with Facebook. The researchers decided to intentionally manipulate the news feed of millions of users to see if they could make their users happier or sadder by selecting more positive or negative content

---

[10] Adam D. I. Kramer, Jamie E. Guillory and Jeffrey T. Hancock, "Experimental evidence of massive-scale emotional contagion through social networks," *Proceedings of the National Academy of Sciences*, 111 (24) 8788-8790, June 2014, accessed online November 17, 2017

posted on individuals news feeds. The research did show that presenting more negative material on a news feed did produce more negative emotions and posts, whereas more positive posts elicited more subsequent positive posts showing for the first time that emotional contagion works across social media.

A number of researchers and pundits criticized Facebook for undertaking research on its users without disclosing the program. Usually, any human research needs to be reviewed by an ethics committee, and the subjects need to consent to the investigation. However, as a private company, Facebook claims to not be bound by the same rules as public institutions like a university and that Facebook users consented to the use of their data by Facebook for research when they signed up with the company. The emotional contagion research was not the first or last time Facebook performed research on their vast social network. In 2012, Facebook published a paper in the prestigious journal *Nature* detailing an experiment they ran to understand how to influence voter turnout in the 2010 United States midterm elections. The paper titled, "A 61-Million-Person Experiment in Social Influence and Political Mobilization," published in 2012 details a simple experiment.[11] In the experiment, Facebook on election day posted one of two messages at the top the News Feed of every eligible US voter on Facebook, which was over 61 million people at the time. One message called the *The Informational Message* noted that it is election day, provided a link to get poll information, and a button

---

[11] Robert M. Bond, Christopher J. Fariss, Jason J. Jones, Adam D. I. Kramer, Cameron Marlow, Jaime E. Settle and James H. Fowler, "A 61-million-person experiment in social influence and political

to click that says "I voted." The other message called the *Social Message* had the same information and button but included the pictures of any of a Facebook user's friends who have already clicked the *I voted* button. The simple addition of the pictures of friends who had already voted nudged an additional 340,000 US citizens to vote in that election. The authors suggest that a single message on Facebook on one single day drove a 0.6% growth in voter turnout between 2006 and 2010. Additional research done by Facebook on its users includes a study done by Facebook to understand reader self-censorship. Anya Zhukova reported on *MUO* that Facebook analyzed the comments and posts that people write but do not submit. (*MUO*, 2017) Facebook found that 71% of its users' self-censor. In other words, an overwhelming majority write things that they feel but do not ever post. Moreover, it is good to know that Facebook saves everything. *Yes, even when you don't send it, it gets recorded.*

Facebook commands a dominant position as one of the world's largest social networking providers. Setting aside, for now, the ethics of experimenting on their users, Facebook has found through their research that they can have a measurable impact on society by manipulating what information they present to their users. The experiment with voter turnout pushed 340,000 voters to the polls that would not have otherwise participated. Considering that close elections have been decided by fewer votes than that, it is essential to recognize the power and reach of Facebook. From the beginning, Facebook has used increasingly sophisticated computer programs to sculpt what individuals see on their news feeds and coupled with the demonstrated effects of emotional contagion; people should exercise some skepticism about what they see on Facebook. A study by Pew Research Foundation recently demonstrated that 40% of adults in the United States get their news from Facebook. Considering that Facebook's algorithms strive to keep your attention by showing you what they think you will like and their parallel motivation to have you click on advertisements, consider going outside the world of Facebook for entertainment, social contact, and news to open your eyes to a different reality than the Facebook world.

Digital leash, or *digital handcuffs,* refer to the consuming effects of multiple devices such as smartphones, tablets, as well as applications from Facebook, Twitter, online dating, and Instagram that hold our attention and focus in a digital world. That digital world stands in contrast to the physical world we inhabit. Elena Bezzubova in *Psychology Today* writes, "We become digital creatures, inhabitants of the new cyber world. At the same time, we remain creatures of the old material world."[12] A term called *digital depersonalization* describes the loss of a sense of self between the many versions of who you project yourself to be online and that person you see reflected in a storefront window or the rearview mirror of your car. With so many projections on different platforms from Instagram to Snapchat, it makes you question who you really are. Dealing with digital depersonalization and avoiding that fine line between infatuation and

---

[12] Elena Bezzubova, "Digital Depersonalization," *Psychology Today,* January, 26, 2018, accessed online March 13, 2018

addiction will allow you to profit from the new digital worlds that entertain us and keep us connected to family and friends. Bryan E. Robinson Ph.D. suggests in "Invasion of the Balance Snatchers" that consciously unplugging from our devices at regular intervals and setting clear boundaries such as no smartphones in bed help maintain a healthy relationship with our digital world. That may sound like an obvious thing to do, but a survey of 1,700 cellphone users by the mobile telephone service provider iPass, Inc. revealed that "61 percent of mobile professionals said that Wi-Fi was 'impossible' or 'very difficult' to give up — more than for sex (58 percent), junk food (42 percent), smoking (41 percent), alcohol (33 percent), or drugs (31 percent)," and "7 percent admitted to checking their smartphone during sex, 72 percent from the toilet, and 11 percent during a funeral."[13]

Learning to profit at a personal level from artificial intelligence depends in part on understanding how such new, disruptive technology works. Knowing how it works helps to identify places where it provides an advantage in your personal life such as dating and managing a busy life. Given the strength of artificial intelligence in performing certain tasks such as data analysis, scheduling, and predictions, we should endeavor to cultivate the skills that artificial intelligence does not do well such as empathy, creativity, and collaboration. Moreover, insights into how you can use artificial intelligence and robotics to your advantage also provides clarity on how to cope with such a new technology without being consumed by it. Connectivity

---

[13] "Two-thirds of Mobile Phone Professionals Feel Anxious Without Wi-Fi, Says iPass Report," www.ipass.com, November 28, 2017

in a virtual world allows people to stay in touch with friends and family like never before, which is a great benefit of these AI-driven tools; however, taking steps to remain conscious of the division between our virtual selves and our physical selves will help us cope with the growing influence of artificial intelligence in our personal lives.

Part 3

Protection from Artificial Intelligence

Chapter 8
Commercial and Professional Protection from AI

To protect is to prevent harm or injury to self, others, things, and places. In the final section of the book, we will move from understanding artificial intelligence and how to benefit from it to a look at how to protect yourself from AI. All new technologies come with the good and the bad. Just as people learned the joy and freedom that the automobile brought, we also learned of the danger involved in tragic car crashes that came along with the new technology. In a similar manner, artificial intelligence will bring opportunity like never before as well as new dangers and challenges in the wake of such disruptive technology. By looking at how artificial intelligence will change our lives, we can prepare and protect ourselves at the professional and commercial level, as a society, and as individuals in our private lives.

Times change, and they appear to be changing at a more rapid pace than ever before. Foster and Kaplan noted in an Innosight report that as of 2012 the average lifespan of a company in the S&P 500 Index had declined 70% since 1958 from 68 years to only 18 years.[1] Not only do companies not live as long as they used to, but the makeup of the S&P 500 also keeps changing. By 2011, companies such as Kodak, Texaco, and Sears had left the stock index of top companies while new ones such as Netflix, eBay, and Google had joined. Just 89 companies remain in the S&P 500 since its inception in 1957. Given the decreasing lifespan of corporations, it follows that even fewer will survive the next 50 years.

The rise and fall of corporations worries people because many people depend on these employers to help provide for themselves and their families. The headlines of papers, the subjects of books, and the conversations of talk shows of late have latched on to the predictions of catastrophic job loss across many sectors of society due to automation stemming from artificial intelligence. In a report from the prestigious management consulting group McKinsey titled, "A Future That Works: Automation, Employment, and Productivity," the authors predict that around 50% of current job activities will be automated by 2050.[2] In another report titled "Workforce of the Future" from consultancy firm PwC, formed in 1998 from the merger of Price Waterhouse and Coopers & Lybrand,

---

[1] "Creative Destruction Whips Through Corporate America," Innosight Report, Winter 2012

[2] "A Future That Works: Automation, Employment and Automation," McKinsey Global Institute, January 2017. Accessed as a pdf online October 30, 2017

the analysts predict that 40% of U.S. jobs are at risk of replacement by 2030.[3]

If you are reading this book, you must be asking yourself whether artificial intelligence is threatening your job. Will a machine replace my job? People have been asking themselves this same question ever since the introduction of the machine. Famously, in the early 19th century, the English textile workers in Northeastern England known as the Luddites smashed power looms and stocking makers to confront the loss of employment to less skilled workers. The new machines introduced in the Industrial Revolution replaced the skilled labor of weavers. The Luddite movement culminated in a region-wide rebellion that ultimately drew in the military to suppress the destruction of property. The tension between society and technology did not end with the Luddites but rather has ebbed and flowed even to today. The *International Center for Technology Assessment*, a think tank established in the 1990s, created a neo-Luddite group called the Jacques Ellul Society. Ellul was an anti-technology professor who taught at Bordeaux University in France. Kirkpatrick Sale describes the society's goals in an article titled, "America's New Luddites." He writes, "In the words of a manifesto for the Second Luddite Congress which took place in Ohio last April [1998], it is 'a leaderless movement of passive resistance to consumerism and the increasingly bizarre and frightening technologies of the Computer Age.'"[4] Workers watching their jobs

---

[3] "Workforce of the future—the competing forces shaping 2030," PwC, 2017, accessed online pdf October 30, 2017

[4] Kirkpatrick Sale, "America's new Luddites," Le Monde Diplomatique, February 1997, http://mondediplo.com/1997/02/20luddites, accessed online November 12, 2016

made obsolete by automation see the direct threat it poses to their sense of worth and the ability to support themselves and their families. Moreover, beyond jobs, some members of the Ellul Society simply do not like the world that technology is creating. Furthermore, their worries stem from their observation of the relentless march of automation in the manufacturing sector alone.

Manufacturing in some cases has been almost entirely automated. Robotic manufacturing has progressed to the *lights-out* status. Lights-out refers to the manufacturing process that requires no human intervention. The factory can run with no lights on because the robots do not need to see what they are doing, and no people need to mind the manufacturing process. Imagine the eerie environment of robots building automobiles in a vast dark factory only lit by the bright flashes and flying sparks of arc welders connecting pieces of steel. In Japan, the robotics company FANUC has been running a lights-out manufacturing facility that has no human labor since 2001. In Japan, the FANUC factory uses lights out robotics to produce more robots. Fully automated production may reduce the time and cost needed to manufacture some goods. According to an article in *The Economist* titled, "Business with the Lights Out," the author describes how the growing wage pressure and intricacy of electronics manufacturing have drawn Philips Electronics to use lights-out manufacturing for its electric razor production.[5] In 2016, Foxconn, the massive Chinese electronics manufacturer who makes iPhones for Apple, announced that it was laying off

---

[5] Pete Swabey, "Business with the Lights Out," *The Economist*, Junw 15, 2015, accessed online January 4, 2017

60,000 workers from its assembly line and replacing them with robotic manufacturing. Such a massive shift in employment from humans to robots strikes fear in the hearts of many who depend on these employers for their livelihood.

Throughout history, people have experienced massive shifts in culture and labor. Today, as in the past, jobs once performed by people continue to be replaced by automation. One stunning example occurred in agriculture in the United States. An FDA report indicates the staggering changes in US Agriculture over the past century.[6] In 1900, 41% of the US population was employed in agriculture working on small diversified farms with the help of millions of work animals while today large highly mechanized farms employ only about 2% of the population. The massive decrease in farm jobs does not mean full employment either. Many farm operators need to take off-farm jobs to support themselves. In fact, in 1930, about one-third of farm operators earned off-farm income. In contrast, by the year 2000, 93% of farm operators earned off-farm income. The off-farm income includes local manufacturing and service jobs as well as jobs in metropolitan areas. It is inconceivable how much life has changed.

How to Know if My Work Is Automatable

Andrew McAfee noted in his book *Enterprise 2.0* that the jobs most at risk of automation are those that are repetitive, require little creativity, and involves no

---

[6] FDA report http://www.ers.usda.gov/media/259572/eib3_1_.pdf
Accessed on line November 3, 2016

manual dexterity.[7] If you think of a list of types of employment organized from purely manual labor such as gardener or hairdresser through cashier to accountant and engineer ending with the most creative positions such as artist, architect, or scientist, the safest jobs appear to lie at the beginning and end of that list. For example, landscaping requires manual dexterity and aesthetic capabilities that robotics cannot come close to doing for now. The same goes for hairdressing. However, jobs in the middle such as cashier, bank teller, and accountant are increasingly being replaced by automation. 34.5% of Americans use TurboTax or similar software to prepare and file their taxes instead of either doing their taxes by filling out their forms or going to a tax specialist like HR Block or an accountant.[8] Ask yourself how often you physically go to a bank rather than banking online or using an ATM. When I was little, the only way to check out of a grocery store was through a line with a cashier at a manual cash register with a bag boy who always offered to help take your groceries to your car and even load them up for you. There are so many examples of jobs that have just disappeared over the past several decades. When I was young, I remember going with my dad to the drug store and stopping at the large photo counter with the wall of little yellow Kodak and green Fuji film boxes behind it, and there was always a salesman there to assist you in film purchasing and development. Today, due to digital cameras, the film market is virtually gone, and the home

---

[7] Andrew McAfee, *Enterprise 2.0*, Harvard Business Press, 2009, Boston, Massachusetts

[8] Elyssa Kirkham, "43% of Americans File Taxes From the Comfort of Their Home," Survey Finds, GOBankingRates, January 25, 2016, accessed online March 12, 2017

printer or automated printing kiosk replaced the offsite printing market. Kodak, Inc. although instrumental in developing digital photography, holding the basic patents in digital image capture and processing, missed the digital boat so to speak. In fact, Kodak developed the first digital camera in 1975; however, management decided not to pursue it further for fear that it would negatively impact their film industry. They were right. By the late 1990s, Kodak's photographic film sales due to digital competition crashed, leading the once globally dominant company to file Chapter 11 bankruptcy in 2012.

In 2013, Frey and Osborn of Oxford University released research they performed to determine which jobs were most likely to be impacted by computerization.[9] Their work looked at hundreds of jobs and the cognitive and physical skills required to do that work. The authors, employing mathematical models, classified hundreds of jobs by the likelihood of them being replaced by computers and robots. They broke many jobs into several categories. Some work requires problem-solving such as sorting or a series of manufacturing steps while other jobs take advantage of human dexterity such as mechanics, surgeons, or plumbers. The findings mirror what has been the trend through the Industrial Revolution. Automation threatens to replace repetitive—even complex—repetitive jobs.

Take, for example, a hairdresser. The hairdresser relies on physical ability to wash, cut, and style a client's

---

[9] Carl Benedikt Frey and Michael A. Osborne, "The Future of Employment: How Susceptible are Jobs to Computerization?" A Working Paper, The Oxford Martin School, September 17, 2013.

hair. But there is more to the hairdresser than that. Hairstyle is an aesthetic, personal expression that the hairdresser needs to understand and interpret based on the customer's wishes and limited by the type of hair the customer has. If the customer comes in with straight red hair but wants to have curly dark hair, the hairdresser needs to work with the customer to negotiate the best way to achieve a happy result. The hairdresser may deploy a variety of techniques and products to achieve a certain hairstyle. A good hairdresser using experience and their sense of style will be able to counsel the client as to how this may look given their other features such as skin tone and personal style. Hairdressing requires multiple skills from interpersonal communication to physical dexterity and a level of artistry too.

Given the multiple skills needed to be a hairdresser, it makes sense to look at all the things you do in your job and ask the question: "Can a computer do what I do now or in the near future?" The research of Frey and Osborn looked at hundreds of jobs by the skills associated with those positions and ranked the barrier to computerization based on three primary categories: problem-solving, physical skills, and social ability. We will look at each of these in more detail to help you evaluate your work.

Problem Solving

We have looked throughout the book at applications of artificial intelligence regarding problem-solving, including the sorting of objects, deciding how to navigate an autonomous vehicle, and suggesting treatment regimens for patients suffering from certain types of cancer. Certain types of problem-solving will be automated. However, that statement needs qualification with the fact that machine learning still needs massive training sets to learn how to do a specific task, and those data sets may not yet exist for many functions. Repetitive problem solving with a limited set of standard answers will be easier to automate such as tax preparation. In the past, many individuals with the help of tax forms and guidebooks from the government would prepare their own taxes. Over the past two decades though, computer-assisted tax preparation with TurboTax or other software has significantly impacted tax preparation in the US. In the most recent data comparing the years 2013 and 2014, the IRS reported a 5.8% increase in the number of individuals filing electronically and a 2.2% decrease in tax professionals filing electronically.[10] For the bulk of tax filings, the process follows a relatively simple course and does not require complex reasoning or creativity. Without the need for creativity or aesthetics, tax preparation, and accounting more generally, has given way to computerization and eventually to automation.

Prediction is a type of problem-solving that we looked at throughout the book and, in some cases, will be handled by machines instead of people. Logistics teams traditionally depended on past sales records and inventory information to make predictions about future sales, but artificial intelligence systems can additionally consume large amounts of data about trends and sentiment from the internet and other sources to enable its predictions. Note the earlier

---

[10] "More Taxpayers Filing from Home Computers in 2014, Many Taxpayers Eligible to Use Free File," IRS.gov, IR-2014-28, March 13, 2014, accessed online on 7/1/2016

success of artificial intelligence in predicting election outcomes. In an article titled, "The Rise of AI in Retail and Logistics," the author notes, "Using AI, German online retailer, *Otto*, predicts with 90 percent accuracy what will be sold within the next thirty days and has reduced the amount of surplus stock it holds by a fifth. It has also reduced the number of returns by over two million products a year."[11]   It is fair to point out an insightful comment by the neuroscientist and artificial intelligence entrepreneur Gary Markus that we should remember that *artificial intelligence still needs massive amounts of data from which to learn.* He parodied Andrew Ng's quote about if a person needs less than a second to make a decision, then a machine can do it too. "If a typical person can do a mental task with less than one second of thought, and we can gather an enormous amount of directly relevant data, we have a fighting chance—so long as the test data are not too terribly different from the training data, and the domain does not change too much over time."[12] The point being that not all problem-solving situations apply to artificial intelligence. So, take note of the type of work you are doing and examine it closely to see if the kinds of decisions you are being asked to make depend on abundant and readily available information. Abundant and readily available information might apply in a quality control position separating good products from defective ones. The manufacturer may have many examples of good products and bad products

---

[11] Uwe Hennig, "The Rise of AI in Retail and Logistics," *AI Business*, July 20, 2017, accessed online November 1, 2017
[12] Steven Max Patterson, "What AI can and cannot do today," *Network World*, March 30, 2017, accessed online October 26, 2017

with which to train a machine.

Physical Skills

First, let us look at physical skills from three different angles: use of fingers, complex hand movement, and working under challenging spaces or conditions. Does your job require you to use that exceptional human skill of doing precise and intricate work with your fingers? Manipulating small pieces in manufacturing such as luxury watches at Rolex is still done by hand. Handcrafted quality still rings true in many cases as the highest quality of all. Other jobs too, such as a surgeon, require precise finger control in making incisions and sewing up patients. Robot-assisted surgery has evolved rapidly over the past 20 years. In cardiovascular surgery, neurosurgery, gastrointestinal, and orthopedics, robot-assisted surgery continues to grow. ROBODOC came into use around 1992 for bone cutting, and newer robotics assist in knee and hip replacement. The past several decades witnessed the development of many areas of surgical intervention with robotic assistants. Using systems of pulleys, micromanipulators, or even computer-guided systems, surgeons have gained high precision and in many cases better outcomes than before. Once regarded as safe from computer replacement, surgeons too will see the effects of automation on their careers. Perhaps automation will not replace surgeons, but robots will be an extension and improvement of their craft.

Many other jobs require finger dexterity such as the dental hygienist. Although the routine task of cleaning and polishing teeth is susceptible to automation, the variation in mouths from patient to patient does present a more complex environment for a robot coupled with the fact that the patient is almost always awake, providing a moving target for the robot, making this procedure an uphill battle to automation. In a similar vein, occupational or physical therapist falls under the category of challenging to automate due to the variety of patients that a therapist must work with, the different ailments presented, and various body types. According to "5 Big Advantages of Becoming an Occupational Therapy Assistant," occupational therapy offers a wide range of settings from hospitals to schools and public and private residences.[13] Additionally, occupational therapy also provides a wide range of specialization from elderly care, sports medicine, occupational injuries, and more. The point is that the variety, complexity, and human interaction required in this growing field shields it from automation.

Beyond finger dexterity, controlled hand movement in changing and unpredictable environments favors human labor over automation for now and the near future. A simple example would be a landscaper. Moving through a landscape with uneven terrain, encountering plants of various heights, using different tools such as sheers, mowers, leaf blowers, and edgers combined with the skills of tree pruning and pest control makes this type of work very safe from automation. In particular, two things from the assessment of a landscaper's job that help protect it from automation are complex terrain, which robots are getting better a negotiating but still not commercially viable, and the second aspect is a universal protectant—aesthetics.

---

[13] "5 Big Advantages of Becoming an Occupational Therapy Assistant," trade-schools.net, October 26, 2017, accessed online August 21, 2017

Knowing what is aesthetically pleasing for a given space and context provides a significant challenge to the automation of landscaping. Understanding how to trim certain shrubs so that they look nice along a walkway or deciding what type and how much ground cover would look best under your stand of dogwoods requires *taste*. However, one particular aspect of landscaping has had robot competition since the mid-1990s, robotic lawn mowers that work similarly to the familiar Roomba autonomous indoor vacuum. Husqvarna was the first to offer a robotic lawnmower commercially. It works by leaving its docking station where it charges its batteries and then randomly running around your lawn for a prescribed amount of time before it navigates back to its docking station. A low voltage wire buried around the edge of your lawn tells the robot-mower not to go any further, keeping it from mowing your flowers or wandering the street or your neighbor's flowerbed. Priced in the thousands of dollars with newer brands launching lower cost alternatives, the robot mower may take over the Saturday morning job of mowing the grass, but mowing the lawn is only a part of the landscaper's job of yard beautification and maintenance.

Many jobs require *manual complexity*, which means having to move your hands into different and unpredictable positions. Consider the variation in tasks and altitude that a construction worker faces on the job, which explains why that type of work remains highly resistant to automation. Construction work, including home building, industrial building, and public works such as roads, bridges, power plants, and airports, requires a massive variety of physical and mental challenges. Something as simple as climbing a ladder and then walking across scaffolding, although scary to some people, poses a considerable challenge to robots today.

Very few jobs appear to be utterly immune to automation, but construction, compared to factory work with its many robotic applications, has been a big exception. Construction has always relied on tools from the simple hammer, saw, and drill to more complex devices that amplify the work one person can do such as cranes, backhoes, and bulldozers. Most tools today are available in a powered version from saws to nail guns to screwdrivers, and power tools have significantly reduced the number of construction workers needed to build houses and office towers. But construction work seldom happens indoors in a controlled environment—the place that robots are most effective. The construction site will endure changes in weather such as wind and temperature. Take for example the wind effects on skyscrapers, and imagine being the construction worker at the top of the world's tallest building. The 2,717-foot-tall Burj Khalifa in Dubai stands over one thousand feet taller than the next tallest building in the world. At that massive height, over 27 football fields high, the wind pressure sways the top up to six feet back and forth. Construction workers possess the essential ability to adapt to varying conditions such as wind-sway or temperature changes. Human adaptability, which has been very difficult to build into robots, has traditionally insulated construction from automation.

Although construction, for the most part, has resisted automation, that too may change. A robotics company out of New York called Construction Robotics has introduced the first semi-automated bricklayer called SAM. SAM for *Semi-Automated Mason* works as a brick layer's assistant laying bricks at a rate four times faster than a human. SAM has some unique capabilities too, such that SAM can handle if the scaffold holding it up sways in the wind. Also, like a human, SAM adapts to slight deviations from the plan specifications. The semi-automated mason does not replace the mason but makes him 3-5 times more productive as well as reducing work-related injury by taking on 80% of the lifting. Unlike a human, SAM does not do the aesthetic work such as cleaning up grouting errors nor can it handle difficult locations or corners.

Which brings up another physical skill humans possess that opposes automation, and that is the ability to work in irregular spaces. Consider the residential plumber doing repairs in an older house. Often having to work under sinks or having to reach into awkward spaces, the plumber must deal with variable and unpredictable terrain with each different client's home. We can see a pattern emerging here for jobs that are more insulated from automation—irregular physical space. People are flexible and adaptable to changes in the environment, and that is an essential capability in plumbing. Of course, improvements in design and manufacturing will eventually produce more versatile and flexible tools and even automated assistants. Physical dexterity and adaptability continue to be a barrier to automation across many jobs. When you think about your current employment and its future, ask yourself what physical requirements your job demands and honestly assess if it is complicated or varied enough not to be automated.

Social skills

Researchers have noted that beyond complex

physical skills that require coordination and working in irregular terrain, and an aesthetic sensibility, the third category of human capabilities to stand in the way of automation is social skills. Social skills include reading other people's emotions, understanding social situations, convincing others to do something, brokering deals, and compassion for others. The social, emotional space has been a significant area of research in artificial intelligence and robotics for many years. Consider jobs such as social work, counseling, physical therapy, spiritual leadership, healthcare, child and elderly care, and the like that require social skills. The ability to understand non-verbal cues, read emotional states, and tailor one's responses and actions poses a significant hurdle to automation.

Try to imagine a robotic priest. What if the machine was able to listen carefully to your problems, have a deep recall of holy scripture, and be able to suggest a combination of practical approaches to your problems coupled with readings from the Bible? Would that robotic *priest* be of help to a congregation? It may not be as farfetched as one might think.

Art often shows us glimpses of the future, and the Spike Jonze movie *Her* offers a computer alternative to inter-human love. *Her* depicts a lonely man falling in love with the world's first artificially intelligent operating system. Through a series of questions, the operating system develops a unique relationship with the main character named Theodore Twombly. Theodore names the AI companion Samantha, and it has the voice of Scarlett Johansson. The movie follows the growing relationship between Samantha and Theodore as the AI grows and learns, and Theodore navigates his broken love life. The concept feels implausible, but perhaps it is not. Today, people are comfortable asking questions of their computer or smartphone through services such as Apple's Siri or Microsoft's Cortana.

Social skills most at risk for automation are those that are repeatable and predictable just as is the case with mental and physical abilities mentioned in previous sections. In an article in *Business Insider*, "Chatbots are Revolutionizing Customer Service," the author suggests that artificial intelligence-powered chatbots will soon take over customer interactions with businesses. "Today's users want to be empowered to help themselves and get things done instantly without assisted service. Gartner agrees and predicts that by 2020, customers will manage 85% of their relationship with an enterprise without interacting with a human."[14] With advances in voice recognition and language processing, chatbots may be the frontline of interaction between companies and their customers; however, delivering a satisfying customer experience may prove more elusive than Gartner predicts. Take for example the role of the bank teller that analysts thought would be made obsolete by the first automatic bank tellers (ATMs) and then online banking. However, in contrast to expectations, bank tellers continue to be an important part of the banking experience with only a slight decline since the peak in the 1980s. "The Bureau of Labor Statistics estimates a nearly 8% decrease in their numbers from 2014 to 2024, from 520,000 to 480,000."[15] Bank tellers have seen their roles changing over the years from the traditional transactions such as

---

[14] Don Davidge, "Chatbots are Revolutionizing Customer Service," *Answer Dash*, November 17, 2016, accessed online October 31, 2017
[15] Amber Murakami-Fester, "Why bank tellers won't become extinct any

cashing checks and dispensing money to assisting with loan applications and other financial services. The point being, the social skills needed for better customer satisfaction and more successful banking needs the human connection.

Studies show that advances in deep learning and machine learning techniques now regularly outperform humans in precision for specific tasks such as facial recognition. One particular case from the legal profession is the correct identification of documents relevant to a legal case. Apparently, the sheer volume of data and rising legal costs creates the opportunity for artificial intelligence to impact the practice of law, especially in document discovery. However, law naturally encompasses much more than document discovery. The authors Dana Remus, Professor of Law, University of North Carolina School of Law and Frank Levy Professor Emeritus, Department of Urban Studies and Planning, Massachusetts Institute of Technology, published an analysis of the potential effects of artificial intelligence on the legal profession. In the paper titled, "Can Robots Be Lawyers? Computers, Lawyers and the Practice of Law," they study the potential employment effects of automation for lawyers based on an analysis of various lawyering activities.[16] The authors used data from a legal billing firm, Sky Analytics, to determine how lawyers allocate their time, and the authors examined the nature of the various activities in light

---

time soon," *USA Today*, March 27, 2017, accessed online November 1, 2017
[16] Dana Remus, and Frank S. Levy, "Can Robots Be Lawyers? Computers, Lawyers, and the Practice of Law," (November 27, 2016). Available at SSRN: https://ssrn.com/abstract=2701092, accessed on line February 19, 2017

of the potential for automation. They looked at which lawyer activities lend themselves to automation and those that require emotional IQ and higher order understanding—two qualities that are beyond the reach of computers for at least the next decade. They grouped the billable legal activities into thirteen work categories. Four representative categories are listed here with the percent of time spent in each activity in the largest law firms, and the authors' estimation of the effect of automation on legal jobs:

◊ Document Review 4.1% (very vulnerable to automation)

◊ Document Drafting 5.0% (vulnerable to automation)

◊ Legal Writing 11.4% (Resistant to automation)

◊ Court Appearances and Preparation 13.9% (very resistant to automation)

Remus and Levy looked at some other activities not listed above, and the reader should look at their publication for a complete list. Here we examine a subset of lawyering activities that illuminate which activities appear most vulnerable to automation and for comparison those determined to be more immune. In light of the capabilities demonstrated by ROSS and the training of Watson to perform document review, it makes sense that this aspect of lawyering would be the most likely to be impacted. Although ROSS came first, expect the space to get crowded soon, as other companies like Chicago based NexLP bring similar products to market. NexLP uses machine learning also to do document review. They claim that they can find the narrative in massive amounts of data. For example, in securities fraud cases, NexLP summarizes conversations or email contents and connects these to fluctuations in stock prices. After establishing the narrative, NexLP pulls the relevant documents for the legal team. More advances in document review appear inevitable, making this space less and less safe among lawyering activities.

In legal practice, document drafting involves the drawing up of wills, contracts, and deeds. The next time you have to sign a contract for a mortgage or click *agree* to the terms of use for some software, you should take notice of the long and precise language in it. And if you asked the attorney if she wrote all of the verbiages in the mortgage, she would have said, *no*, that the contract comes from a template. Attorneys have used templates for many years. Remus and Levy point out in their article that lawyers already access most templates through computers already; so there is not much room for innovation and not much impact on lawyer employment. Except they point out that companies like LegalZoom provide templates for documents such as wills and divorce filings *without a lawyer present*. Certainly, LegalZoom and other direct-to-consumer services will impact small legal practices that depend on the drafting of wills, trademark filing, real estate lease, and corporate name change to name few for their regular practices.

Legal writing, on the other hand, requires significantly more interpretation than document drafting and will constitute a remarkably higher challenge to automation for some time. Lawyers, judges, and lawmakers use legal writing for the drafting of resolutions, judicial opinions, rights, and responsibilities. Some of this very technical writing may seem robotic, but some of the best legal writing not only clarifies the law but can be great reading on its own. Take, for example, the clarity and power of the writing of the late Supreme Court Justice of the United States, Oliver Wendell Holmes, in his famous dissent written for *Abrams v. the United* States decided in 1919.

"But when men have realized that time has upset many fighting faiths, they may come to believe even more than they believe the very foundations of their own conduct that the ultimate good desired is better reached by free trade in ideas—that the best test of truth is the power of the thought to get itself accepted in the competition of the market, and that truth is the only ground upon which their wishes safely can be carried out."[17]

Holmes pivotal opinion marked a change in the nation's attitude toward free speech and is subsequently cited in many cases in law. Global Freedom of Expression is an online legal database from Columbia University focused on law that relates to freedom of expression. The database claims that 450 legal cases have cited Holmes' dissenting opinion since its publication in 1919.[18] The remarkable writing of Holmes has inspired people far beyond the bench. The point is not to just recognize great legal writing but to highlight the creative, multidimensional aspects of legal writing. In essence, legal writing requires theme development, the use of compelling examples, and prose that will illuminate and persuade. If legal writing were structured or formulaic, one could see how machines

---

[17] *Abrams v. United States*, 250 US 616
[18] *Abrams v. United States*, Global Freedom of Expression, Columbia University, Accessed on February 24, 2017

How to Profit and Protect Yourself From AI

could take over the task of legal writing. Recall that machine learning does very well with structured and repeated elements. Some legal writing may fall into this realm for certain boilerplate-type writing, but all in all, the command of language and emotional intelligence needed for legal writing goes beyond the current and near future capabilities of artificial intelligence, which protects this aspect of lawyering from significant impact by artificial intelligence.

By a similar token, Remus and Levy consider court appearances and preparation to be far from making the endangered species list. The presentation of arguments to the jury, cross-examining a witness, persuading a judge, and characterizing a client all require a wide range of skills. Shepherd and Cherrick sum up this point beautifully in their essay "Advocacy and Emotion."[19]

If you have any doubt about the central role of emotion in the art of advocacy, refer to Aristotle and Quintilian, two classic teachers of oratory and rhetoric. They taught that a persuasive written or oral presentation contains three elements:

◊ ETHOS: The ethics, integrity, and character of the advocate

◊ PATHOS: The emotions that the advocate instills in the audience

◊ LOGOS: The logic or reason that supports the advocate's argument

A modern litigator's success depends on applying

---

[19] John C. Shepherd and Jordan B. Cherrick, "Advocacy and Emotion," *Journal of the Association of Legal Writing Directors*, Volume 3, Fall 2006

the wisdom of these ancient philosophers. Emotion and reason constitute the necessary elements of all persuasive arguments.

The range of emotional and intellectual intelligence needed in the courtroom safeguards court appearance and preparation from automation for years to come. Moreover, the rules governing advocacy and who may present evidence or advocate for a client prevents a robot from representing a client, but there may come a time when Watson attempts to pass the Bar and new legal questions about how can an attorney who lives in the cloud be allowed to approach the bench.

As we have seen, the legal profession today has many categories of activity, and just as in other professions, the more routine tasks are feeling the pressure of automation and rapid encroachment of artificial intelligence. Note, that of the lawyering activities mentioned above, only document review fell under the sharp employment effects category with the rapidly developing AI capabilities of ROSS, NexLP, and more. However, the task of document review only takes about 4% of the average lawyer's time. The wholesale replacement of a lawyer remains the subject of science fiction, but the law firm of today and definitely of tomorrow differ vastly from the traditional law firm. The division of labor will change with automation. Gone will be the army of junior attorneys working 18 hours a day in document review. Instead, the junior attorneys will need to understand the technology they rely on and to focus more on the activities best suited to people such as legal writing, court appearances, negotiation, and communication.

In this section, we have applied, with the help from the work by Remus and Levy, the analysis of various aspects of the work that lawyers do. Popular writing and futurists predict that automation will completely replace many lawyers. Remus and Levy take a more conservative view and conclude only one sector of lawyering will take a significant hit, and that is in document review. They predict an 85% decrease in time lawyers devote to this task. Remember the categories that indicate the potential risk of automation for each of the three main categories—physical, cognitive, and social. Overall, the risk of automation in the legal profession is medium to low except in the case of document review. Notably, there is no real physical aspect to lawyering that is challenging for a robot to perform such as difficult terrain or physical dexterity; however, the requirement of the physical presence of lawyers in the courtroom will protect that aspect of lawyering for years to come. Finally, the social elements of argumentation and working to meet client needs from understanding their desired outcome to relating to jurors and judges in the courtroom will protect legal jobs for the foreseeable future.

Case study—Analyst

Our jobs face the challenge of automation, stemming from machine learning and advances in robotics. In this chapter, we looked at three broad areas of human capability that pose considerable challenges to computerization: physical abilities, cognitive skills, and social skills. Some examples indicate that machines surpass humans and in other cases, machines still face an uphill battle. However, as Ray Kurzweil and other like-minded futurists suggest, technology continues to accelerate, thereby introducing challenges to our lives and livelihoods at a pace not ever seen before in our history.

In light of our rapidly advancing technology, everyone should carefully examine their jobs to see how potentially vulnerable they are to automation. If you are a parent, teacher, guidance counselor, friend, pastor, coach or concerned neighbor, automation must be included in considering a youth's education and career path. You must plot out your goals and plans not only in light of personal fulfillment and potential financial success but carefully consider your current or future work in light of the three areas of human skills that currently resist automation. Sit down and work through all the aspects of your current work or the job you would like to do. Research the types of work you want to do and determine what areas are changing. Be thorough and rigorous. CollegeBoard.org offers many career planning resources online, including their projections of what jobs will be in the highest demand over an eight-year period.[20] CollegeBoard used government economists to estimate where the most jobs will be in 2018, and market research analyst made the list of top ten jobs for individuals with a bachelor's degree. For this chapter, market research analyst will be evaluated by the three categories for risk of automation. To begin this process for any job you want to examine, gather as many descriptions of the work as you can from job description publications and websites such as CollegeBoard and O*Net, or look at the many job offerings found on job sites like Monster. com, CareerBuilder.com, and the like. Gather the

---

[20] "Hottest Careers for College Graduates, Experts Predict Where the Jobs Will Be in 2018," https://bigfuture.collegeboard.org/explore-

job descriptions and start to estimate the likelihood of automation for your job. Also, use the terms and descriptions to research whether there are upcoming or established services available commercially or even free as open source software that would automate the job or part of the job.

The following is the aggregated description of the requirements for a market research analyst compiled from various sources: a market research analyst collects information from the public by different means such as surveys, interviews, social media analytics, then analyzes the information, forms opinions and recommendations, and then communicates the findings in writing or presentations. The work has applications across many sectors from commercial to academic areas to political campaigns to customer satisfaction. O*net details a number of job requirements for market research analyst, including tasks such as devising surveys and collecting data to determine market share, skills needed that include complex problem solving, critical thinking, and active listening. Other noted requirements include tools and technologies that are mainly computers and a wide range of software from survey tools to statistical analysis tools such as MATLAB and STATISTICA to customer management tools and web design. A good market research analyst must possess strong language skills, customer service, and management capabilities. The online job description calls explicitly for strong quantitative skills and persuasive writing capabilities to support findings and assertions from research. The requirements for the high growth field of market

---

careers/careers/hottest-careers-for-college-graduates

research analyst can be broken down into three main categories—physical, cognitive, and social.

The risk of automation for market research analyst in the physical category scores high because there is no specific material quality to the work that would pose a challenge to automation engineers. The only physical demands mentioned in the job descriptions specify frequent travel and long hours. Neither of these would protect a market research analyst work in the physical category.

In the *cognitive category* of automation risk, the requirements for market research analyst presents a variety of skills and qualities that in some cases appear ripe for automation and others that would not be easy to automate in the near term. First, most descriptions of the job call for skills using a wide variety of software tools to gather, analyze, and report data. Many software tools now assist the market research analysts in ways that were not available even a few years ago. Before the advent of the internet and email, running a survey, which many market research analysts do to gather information, consisted of either paper surveys or direct subject contact. The survey data needed to be collected and analyzed by hand. Today, many software tools cover the entire survey process with templates, questions, built-in survey analysis components, even automatic follow up emails sent to non-responders. Simple surveys are easy to automate now, and sentiment analysis across various channels lends itself better to automated AI tools that can track millions of responses in real time across platforms like Twitter. Unlike ever before, we can see in real time the reaction to events across the country or the world such as sentiment to various commercials played during the Super Bowl or the changing views towards a political candidate's speech or response to breaking news. It is important to examine what elements of the market research analyst's job are routine and therefore easily automated and those that are non-routine. Many routine tasks for the market research analyst appear ripe for automation now or in the near future, but the non-routine aspects appear resistant to automation which is why the overall risk of automation is medium in this analysis.

The last category to look at in the example of evaluating the risk of automation for a market research analyst is the *social* aspects of the work. For the social evaluation, remember to look at the job description of the market research analyst for several aspects of social skills, including the ability to read and interpret others' reactions, to bargain, to influence and to work with and help others. The New England College's description of a market research analyst breaks the job into two main categories—quantitative and qualitative. Quantitative work is strongly dependent on statistics to derive useful information from surveys such as market trends from anything such as, "How is America changing its breakfast habits?" to "Will people accept driverless cars?" Statistics help researchers draw conclusions about a large group of people with only having to gather a small amount of information from them. The other type of market research, *qualitative*, relies on in-depth interviews with a small number of people to derive insights into a problem. Conducting in-depth interviews requires strong social intelligence to connect on a personal level. Also, understanding research results and persuading your boss or clients to accept your interpretations and evaluations will draw heavily on your social IQ. Furthermore, the ethical implications of using data such as customer browsing history and online shopping activity demand a robust social intelligence. What data to gather and privacy protection are the responsibility of the market research analyst. Considering the high social skills needed for this job, the risk of automation appears to be low.

Looking at one job in light of its physical, cognitive, and social requirement and ranking the likelihood of automation illustrates a vital technique for looking at your work or what type of work you may want to do in the future. Conduct this exercise from time to time so you can be proactive and informed about the potential obstacle to your career and plan accordingly. Look around; many grocery stores have self-service checkout lanes now. One may think, well I am a cashier and although my job is not physically challenging nor does it require me to do much non-routine thinking or problem solving, I do have a social job. I greet customers and may engage in conversation; this should protect my position from automation. Your social skills may not be enough to protect your job from automation. The US Department of Labor predicts that due to automation and increased online shopping, the growth of cashier jobs will be much slower than most jobs in the United States.[21]

Jobs will be taken or changed with the rise if AI, robotics, and ultimately AGI. It is already happening across the globe. Be proactive about your future and prepare for the changes to come. Look at the many

---

[21] Bureau of Labor Statistics, U.S. Department of Labor, *Occupational Outlook Handbook, 2016-17 Edition*, Cashiers, http://www.bls.gov/ooh/sales/cashiers.htm, accessed online July 17, 2016.

employment prediction websites to see where job growth is going, but especially ask the critical questions outlined above, do the research, and even contact people doing the job you are interested in doing and ask them what aspects of their job they think will be automated soon. On a final note, knowledge worker jobs, or jobs which work with information such as lawyers, pharmacists, and scientists, have steadily increased over the past ten years based on statistics from the US Department of Labor. As routine jobs disappear, new jobs continue to emerge. Look for the growth areas and focus on tasks that still need the physical, cognitive, and social skills that do not lend themselves to automation.

Chapter 9
Protecting Society and Ourselves from AI

Protection, Security, Safety

Humanity stands on the precipice of a new era, unlike anything we have seen before. In a way, we have been wandering along this precipice since the Industrial Revolution was declared over in the 1980s. The Information Age, also called the Digital Age, had dawned, and we as a culture continue to transition to the new Digital Age. The embryonic Information Age continues to evolve and grow. The Industrial Revolution, with its need for skilled labor and the growing mechanization of farming, drew most of the rural population off the farm and into city centers. Employment options diversified radically from the manual labor needed on the farm to manufacturing and service jobs. Diverse manufacturing from heavy industry producing steel for everything from rails for the expanding train system to parts for cars and every other sort of product imaginable required a new skilled labor force. Skilled labor in factories made up only a part of the expanding labor opportunities. Growing industrialization created more wealth and expanded the middle class, creating a new consumer that demanded more clothing, better houses, and a wide variety of consumer options. It follows that each expansion required more labor to manufacture new goods but also required more services. The Industrial Revolution vastly expanded the diversity of employment options available, but the extended technology and hunger for wealth also created incredible abuses in the workplace and the environment.

History contains countless examples of labor abuse. Famously, Upton Sinclair in his novel *The Jungle* published in 1906, exposed abuses in the meat packing industry in Chicago in the early 20th century. *The Jungle* highlights the deplorable, dangerous, and unsanitary conditions the meatpackers worked in as well as demonstrated worker abuse by management. Shortly after its publication, the book caused a massive public outcry and a significant drop in the purchase of American meat both at home and abroad. The novel helped in part to drive the passage of the Pure Food and Drug Act in 1906, which led to the creation of the Food and Drug Administration (FDA). The FDA bears the responsibility of protecting the public from the manufacture, sale, or distribution of adulterated, misbranded, poisonous, or deleterious foods, drugs, medicines, and liquors, and for regulating traffic therein, and for other purposes.[1] The Pure Food and Drug Act forms part of a variety of legislation enacted over the past 100 years governing child labor, environmental protection, occupational health and safety, and more. The waves of new laws and the growth of government emerged to protect the public from dangers evolving from industrialization. In the same way, the dawn of The Digital Age brings with it new challenges to society such as the rapid loss of employment to automation, loss of privacy, the threat of cyber-crime, rules and judgements based on intractable algorithms, and perhaps even machines with artificial general intelligence.

History contains many examples of the new technologies released into the world, and only later did their unintended consequences emerge. In *Silent Spring*, Rachel Carson's seminal book published in 1962, she demonstrated how the mass distribution of the pesticide DDT does not just kill insects. DDT builds up throughout the food chain. For example, a bird may consume worms containing DDT from the soil. Birds eat more of the contaminated worms. The birds may not get sick, but the DDT stays in their bodies and builds up over time. A falcon that preys on smaller worm-eating birds will accumulate even more DDT. Again, the DDT does not kill the falcon, but it interferes with its reproduction by thinning the falcon's egg shells. The shells cannot support the weight of the mother in the nest and break. Many magnificent birds of prey were disappearing in the United States such as the Peregrine Falcon and the Bald Eagle due to DDT

---

[1] Federal Food and Drugs Act Of 1906 (The "Wiley Act"), Public Law Number 59-384, 34 Stat. 768 (1906)

contamination. Rachel Carson, through her book, made the world realize that massive pesticide usage had dire and unintended consequences.

Silent Spring marked the beginning of the environmental movement and ultimately the Clean Water Act and other environmental legislation. In 1973, Herbert Boyer of the University of California and Stanley Cohen of Stanford University ushered in the age of genetic engineering by successfully transferring the DNA from one organism to another in a process called *cloning*. Cloning has fundamentally changed biology, medicine, and agriculture not to mention the very nature of how we view life and man's ability to *play God* by manipulating the very blueprint of life. In the early '70s in the shadow of the rising power of government regulatory agencies such as the FDA and EPA and a burgeoning environmental movement, the pioneers of genetic engineering were aware of the potential for their new technology to harm. The growing public concerns included the fear of altering the environment with genetically modified plants, engineering more virulent diseases that could be used in germ warfare, and the concept of altering evolution in unpredictable ways. To stay ahead of regulators and to appease the growing fear from the public, Boyer, Cohen, and other leading scientists of the day organized a symposium at Asilomar Conference Center near Monterey, California in 1975. The purpose was to gather scientists, ethicists, government officials, and journalists to discuss the potential dangers of genetic engineering and to propose actions that would mitigate these dangers.

Remarkably, in the months leading up to Asilomar, scientists from around the world voluntarily agreed to a moratorium on genetic engineering experimentation until the conference. The concern was for the safety of the scientists performing the experiments and for the risk to people and the environment more generally. Asilomar was a ground-breaking event because it was the first time that scientists recognized the potential danger of their new technology and organized a very public debate to address the very issues. A core accomplishment of the symposium was the drafting and agreement on core safety issues including the proper laboratory procedures for the handling of potentially dangerous biomaterials, experimental design to prevent the accidental spread of modified organisms, and the moratorium on the cloning of genes from pathogenic organisms such as smallpox. The conference included journalists and lawyers and took science out of the laboratory for a close examination. Asilomar lasted three days, and the conclusions and guidelines paved the path for remarkable growth and safety in biotech research, industry, and agriculture in the 40 years since then. The organization of computer scientists along with the government and other disciplines in the Future of Life Institute's "Research Priorities for Robust and Beneficial Artificial Intelligence: an Open Letter" falls into the same category of conscious self-regulation demonstrated by Asilomar. However, it remains to be seen if their guidelines will similarly usher in 40 years of positive growth without a major catastrophe from artificial intelligence and robotics.

Privacy, Privacy, Privacy

Apart from self-regulation by the artificial intelligence research community and industry, the government will undoubtedly be called upon to protect the public from harmful aspects of this emergent technology. The United States Federal Code embodied in over 23,000 pages of law provides among other things the power for the Executive Branch of the government to create the rules necessary for enforcing the laws enacted by Congress. Government regulations will undoubtedly cross paths with AI in many areas, including automobile safety, communications, privacy, and digital property rights. The ever-shrinking cost of digital storage and the pervasive intrusion of digital activity in our lives allows social media, information companies, and the internet of things to create and maintain a digital trail for everyone, for hackers, governments, and businesses, to follow our every move and interaction.

The United States National Highway Traffic Safety Administration (NHTSA) regulates the safety standards of the automotive industry and transportation. Safety and product liability around the automotive industry has a rich history in the courts and government regulation. In many respects, the NHTSA would not exist but for the tireless work of Ralph Nader. The US government was doing little to enhance the safety of automobiles in the 1960s. Since his time at Harvard Law School in the 1950s, Ralph Nader studied the cases of automobile safety. In 1965, he released a hugely popular book called *Unsafe at Any Speed*, which critiqued the safety of automobiles in America, virtually indicting the automobile industry for selling cars with known safety issues such as the Chevrolet Corvair. The public reacted so strongly to *Unsafe at Any Speed* that Congress only a year later after its publication in 1966 passed the United States National Highway Traffic Safety Administration Act. The formation of the

NHTSA ushered in an era of significant improvements in automobile safety that has brought crash-testing, anti-lock brakes, airbags, and many other safety features. It is under the safety umbrella of the NHTSA that one of the most complex of embodiments of artificial intelligence, machine learning, and robots continue to take shape today—driverless cars.

Decision Making and Liability

Decision making by the driverless car will be subject to litigation no matter what decision the vehicle makes. But who will be ultimately liable? Is it the owner of the car? Is it the NHTSA who set the safety standards? The questions of liability in the case of accidents involving driverless cars have frequently been discussed both in the legal literature, mainstream press, and government commissions. The issue of acceptable safety with autonomous vehicles was interestingly phrased in the open letter from the Institute for Life, "Research Priorities for Robust and Beneficial Artificial Intelligence." "If self-driving cars cut the roughly 40,000 annual U.S. traffic fatalities in half, the car makers might get not 20,000 thank you notes, but 20,000 lawsuits."[2]

Although the statement is somewhat absurd because you do not mail thank you notes to your automobile manufacturer every time you safely arrive home, it contains the element of hedging against the

---

[2] Stuart Russell, Daniel Dewey, and Max Tegmark, "Research Priorities for Robust and Beneficial Artificial Intelligence, Association for the Advancement of Artificial Intelligence," futureoflife.org, Winter 2015, accessed on line March 19, 2017

expectation of 100% safety. The NHTSA will have to develop guidelines for safety that will allow *acceptable risk*. Knowing that no system will be perfectly safe, there are existing liability laws that will serve in the first wave of lawsuits once autonomous cars come into commercial use, but the problem stems from the current system which places liability on the owner of the vehicle. The owner pays the insurance and is the one responsible for paying damages if their car is involved in an accident. Moreover, following a petition from Google to regulators, The National Highway Transportation and Safety Administration told Google that the artificial intelligence system that controls its self-driving car will be considered a driver under federal law.[3] Now, what happens if an autonomous vehicle hits another autonomous vehicle? In this case, there is no human driver to accept liability. Automobile companies such as Volvo have claimed that they will take full responsibility for the autonomous cars they produce when they are in self-driving mode.[4] However, responsibility will be challenged. When an accident with an autonomous vehicle happens, and it was the result of a failed system such as the laser radar system that tells the car what lies ahead in the road, who accepts the fault? Will fault be assigned to the laser manufacturer, the car manufacturer, or the software engineers who did not design the system to compensate safely for the failed radar? No matter what direction the law takes, there will undoubtedly be plenty of

[3] Kirsten Korosec, "The Artificial Intelligence in Google's Self-Driving Cars Now Qualifies as Legal Driver," *Fortune*, February 10, 2016, accessed online August 6, 2016
[4] Jim Gorzelany, "Volvo Will Accept Liability for Its Self-Driving Cars," *Fortune*, October 9, 2015, accessed online August 6, 2016

opportunity for prosecution.

Protection from Automation in the Law

The phrase *as sober as a judge* reflects the high expectations we have of judges and their responsibility to apply the law objectively and evenly to anyone in their court. Judges bear a heavy burden to dispense the law and render not just judgment but sentencing as well. Given the awesome responsibility placed on judges and the pivotal role they play in the lives of the plaintiffs and defendants that stand before them, it would seem very unlikely that we would hand the role of judge over to a computer. However, artificial intelligence has already begun to enter the judge's chambers. A system sold to courts called COMPAS (Correctional Offender Management Profiling for Alternative Sanctions) has been developed by Northpointe Inc. to evaluate the risk that a criminal will commit another crime. In other words, their recidivism rate.[5] Using artificial intelligence combined with information gathered from the subject in an interview, criminal record, health history, history of violence, gender, race, and more, COMPAS will provide an estimation of whether a defendant will likely commit another crime and a separate evaluation of whether the defendant will probably commit a violent crime. Courts and prisons use COMPAS to help guide sentencing and inmate management.

The acceptance of artificial intelligence in the legal system has not been entirely smooth. For example,

---

[5] "COMPAS Risk & Need Assessment System," Northpoint, http://www.northpointeinc.com/files/downloads/FAQ_Document.pdf, accessed online September 29, 2017

COMPAS has been in the news for several reasons. ProPublica, a not-for-profit newsgroup based in New York, published an evaluation of COMPAS concluding that the risk assessment is racially biased. In the article titled, "How We Analyzed the COMPAS Recidivism Algorithm" published in May 2016, the authors describe how they analyzed the actual recidivism rate of over 10,000 criminal defendants from Broward County in Florida with their predicted recidivism rate.[6] The authors observed that COMPAS accurately predicted the recidivism rate of white defendants 59% of the time and 63% for black defendants. Moreover, the investigation concluded that COMPAS is overly biased in predicting the likelihood of black defendants reoffending. In the COMPAS system, black defendants were twice as likely to be misclassified as a higher risk of reoffending than other racial groups. The authors' conclusions suggest that the COMPAS algorithm lacks the required impartiality needed for the legal system. COMPAS was also featured in the legal case of *Loomis v. Wisconsin* heard by the Wisconsin Supreme Court. The defendant, Eric Loomis, was convicted and later pled guilty to being the driver in a drive-by shooting. In the sentencing of Loomis, he was characterized as "High risk to the community." The assessment was in part derived from information obtained from COMPAS. The defendant argued that because the algorithm used in his sentencing was proprietary and could not be fully examined by the defense, he was denied due process. He continued that without due process, which

---

[6] Jeff Larson, Surya Mattu, Lauren Kirchner and Julia Angwi, "How We Analyzed the COMPAS Recidivism Algorithm," *ProPublica*, May 23, 2016, accessed online September 27, 2017

is guaranteed under the US Constitution, he should be set free. The Wisconsin Supreme Court determined that the sentence would have been given even without the COMPAS information. Loomis appealed to be heard by the Supreme Court of the United States. The Supreme Court denied the request to hear this case but noted, "Sentencing court's use of actuarial risk assessments raises novel constitutional questions that may merit this Court's attention in a future case."[7] In other words, the court feels that computer-aided sentencing will eventually make its way to the Supreme Court for evaluation.

Judges accept the weighty responsibility of determining the guilt or innocence of defendants and the determination of the sentence for the guilty. Automation in dispute resolution and sentencing has already made its way into court with systems like COMPAS, Modria, and DoNotPay. Automated systems will put pressure on replacing human judges because they are fallible. Take for example the now famous study titled, "Extraneous Factors in Judicial Decisions" in which the authors concluded that judges are more likely to hand down a guilty sentence before a break such as lunch or at the end of the day than in the morning or after lunch.[8] Judges may be fallible, but complex, *proprietary* computer programs should not replace them. Technological enhancements to legal decision making must involve *transparent application* to everyone.

---

[7] Amy Howe, "Loomis v. Wisconsin, Supreme Court of the United States," SCOTUSblog, May 26, 2017, accessed online September 29, 2017

[8] Shai Danzigera, Jonathan Levavb, and Liora Avnaim-Pessoa, "Extraneous Factors in Judicial Decisions," *Proceedings of the National*

Freedom

The freedom of the road has been a rallying cry of American car culture and underpinned automobile advertising since the first automobiles began rolling off of assembly lines over a hundred years ago. The car offers the freedom to travel at one's own pace and to follow the roads of one's choosing. With cars, we even have the choice to go over the speed limit, knowing the possibility of getting a ticket if caught. In an article in *Scientific American*, anthropologist Krystal D'Costa points out that by the 1920s, automobile advertising had shifted from more technical descriptions of cars to associating freedom, fun, and independence as the lure of car ownership.[9] Given America's love of automobiles and the spirit of freedom built into our relationship with cars, the American public may not react well to the suggestions by industry leaders that the self-driving car will replace human-driven vehicles. Leaders, such as Elon Musk, the entrepreneur and founder of Tesla, William Clay Ford Jr, executive chairman of Ford Motor Company, and president of Toyota Motor Corp. Akio Toyoda, profess that self-driving cars are the future. Some pundits argue that people should not be allowed to drive if the driverless cars show a higher safety record than human drivers.[10] An attempt to map out the timeline for the arrival and take over of autonomous vehicles by Johana Bhuiyan suggests that a total fleet replacement in the United States would

---

Academy of Sciences of the United States of America, vol. 108 no. 17, April 26, 2011,

[9] Krystal D'Costa, "Choice, Control, Freedom and Car Ownership," Scientific American, April 22, 2013, accessed online November 5, 2017

[10] Matt Vella, "Why You Shouldn't Be Allowed to Drive," *Time*

take about 15 years once the autonomous car has been perfected, which would put the US using only self-driving cars somewhere around the year 2050.[11]   A great deal must change before the autonomous vehicle takes over, including liability laws, security to prevent these cars from being hacked, and most importantly the public's willingness to give over their freedom to machines.

Personal Protection and Privacy

Throughout the book, we have been learning how artificial intelligence works, its strengths and weaknesses, and how to use them to our advantage as well as how to protect and prepare yourself for changes to come. In previous chapters, we looked at the types of work susceptible to automation—repetitive work in controlled environments and limited decision making. When planning your career, make sure to look at the three categories of skills that resist automation:

◊   Physical

◊   Mental

◊   Emotional

The examples given for jobs that require complex physical motions such as a hairdresser or landscaper resist automation because of their complexity and varying terrain.   Other positions such as plumber, firefighter, and veterinary technician fit this same

Magazine, February 25, 2016, accessed online November 5, 2017
[11] Johana Bhuiyan, "The complete timeline to self-driving cars," *The Verge*, May 16, 2016, accessed online November 5, 2017

criterion and equally resist automation. The second category to evaluate is the mental or problem-solving work that is reduceable to machine decision making such as basic tax preparation or document analysis as we saw before. Jobs that use human judgement and problem solving will resist automation, such as designers, computer programmers (so far computers may be learning some things, but they cannot program themselves yet), construction workers, farm supervisors, and many other roles that deal with changing problems and evaluation. The third category of job type that resists automation requires emotional intelligence that cannot be replaced with a computer such as coaches, teachers, counselors, nurses, and others that need more than merely identifying an emotional state but go beyond to involve empathy and intuition. Evaluating and planning your career should include seeking jobs that cannot be easily automated or that combine the best of human skills with the power of computers (or *centaurs*).

Privacy

In an opening sentence to the article, "Privacy and Human Behavior in the Age of Information," the authors declare, "If this is the age of information, then privacy is the issue of our times."[12] Privacy is considered to be the state of a person or group or their information being free from observation by other people. In the United States, privacy is not explicitly protected in the

---

[12] Alessandro Acquisti, Laura Brandimarte, George Loewenstein, "Privacy and Human Behavior in the Age of Information," *Science*, Vol. 347, Issue 6221, pp. 509-514, January 30, 2015,

United States Constitution but rather is protected by a patchwork of amendments to the constitution. The 1st Amendment protects the privacy of beliefs. The 3rd Amendment protects citizens' homes from occupation by soldiers during peacetime. The 4th Amendment protects individuals from unlawful search and seizure, and the 14th Amendment provides more general protection of privacy. Privacy covers the sovereignty of the body as well as information sensitive to an individual. In the information age, the use and distribution of personal information cover a vast array of data: pictures, health records, what you search on the internet, and sensitive financial information. People need a sense of privacy to feel safe and to have an emotional break from being observed. Moreover, people require the psychological safety from the fear of being embarrassed. In the Information Age, people's private information can be transmitted across the internet at the speed of light and saved on computers around the world never to be private again.

In "Why We Care About Privacy" Michael McFarland of the Markkula Center for Applied Ethics and Santa Clara University lists some reasons to protect our privacy[13]. Invasions of privacy as we define them here are of concern for a number of reasons:

◊ The more widely sensitive information is disseminated, the higher the danger of error, misunderstanding, discrimination, prejudice and other abuses

---

[13] McFarland, Michael. "Why We Care about Privacy." *Markkula Center for Applied Ethics*, Santa Clara University, 1 June 2012, www.scu.edu/ethics/focus-areas/internet-ethics/resources/why-we-care-about-privacy/.

◊ The lack of privacy can inhibit personal development and freedom of thought and expression

◊ It makes it more difficult for individuals to form and manage appropriate relationships

◊ It restricts individuals' autonomy by giving them less control over their lives and in particular less power over the access others have to their lives

◊ It is an affront to the dignity of the person

◊ It leaves individuals more vulnerable to the power of government and other large institutions

The subject of privacy in the Information Age could quickly take up several volumes. In light of AI, we will focus on how to protect yourself from the growing use of artificial intelligence to harvest and analyze your information and behavior. Privacy protection applies both online when you use your computer, smartphone, smartwatch, or tablet and *offline* when you engage smart systems like Amazon's Echo or Google Home.

Machine learning helps tech companies make sense of the massive amount of data people generate merely by browsing, shopping online, and using their phones, not to mention all the additional data now provided by internet-enabled objects. OnStar by General Motors (GM) offers many features such as GPS-enabled turn by turn navigation, emergency services, and mechanical diagnostics for your vehicle using cell phone technology to connect your car to the manufacturer. Data from OnStar flows from cars back to GM. The service is sold as a safety and convenience feature, but it also provides a wealth of data back to the car manufacturer. It is not clear about all the ways that GM uses your information, but based on a recent announcement from the insurance and finance arm of General Motors, known as GMAC, they definitely use your data. GMAC automatically sends the miles you traveled each month to your insurance carrier for low mileage discounts. By extension, a simple analysis of your logs by the police could determine all the times in a month you were speeding and fine you accordingly. You are driving your car, and the service you are paying for provides data that could incriminate you. OnStar stands as only one of many digital trails people leave all the time, and such trails pose challenges to your privacy.

Instead of addressing difficult privacy issues, Adele Howe, distinguished professor of computer science at Colorado State University and artificial intelligence researcher, suggested in a quote from *The Globe and Mail*, "We have to get over, at some point, the idea that we have privacy. We don't. We have to redefine what privacy means."[14] Dr. Howe suggests that as technology *continues* to improve our lives and the human condition overall that we make tradeoffs along the way. The significant trade-off, she suggests, is our privacy. Now the concept of privacy differs among cultures across the globe and throughout time. A simple example comes from the changing attitudes around body exposure. The introduction of the bikini by French engineer Louis Réard in 1946 was considered scandalous, and they were even banned in some countries such as Spain, Belgium, and Australia. Today, the bikini appears on

---

[14] Carly Weeks, "Dear valued customer, thank you for giving us all of your personal data," *The Globe and Mail*, Wednesday, June 8, 2011

beaches across North America, Europe, and Australia. Privacy may not be a static thing, but we value it in Western Culture. We are not, as Dr. Howe suggests, going to just roll over and accept the dissolution of our privacy. The last part of this chapter will detail several steps we can take to preserve our privacy in the digital world.

Our actions online leave a digital trail of which many companies take advantage of today. Our digital exhaust has many forms from the searches we perform, websites we visit, products we purchase, what we share on social media, to what we watch on Netflix or other streaming media sites. Moreover, many websites that we visit contain invisible computer codes designed to track our every move. In a competitive market, companies want to spend their marketing dollars most effectively, and they can get a leg up on the competition if they find the most likely customers. Machine learning, as discussed above, helps marketers by building profiles of individuals that predict the most likely customers to buy their products, also known as *targeted marketing*. Targeted marketing thrives on more accurate information about you, but your data also serves criminals committing identity theft and scams. One such fraud reported in *Technology News* involved criminals stealing personal health records. Criminals purchase your health records to file false medical claims for reimbursement, or scam artists may target sick people with fake marketing phone calls selling herbal cures for what is ailing them. To protect yourself online you can manage the *cookies* on your computer, turn off your smart speaker, and install a virtual private network.

Cookies refer to little bits of information stored on your computer by a website you have visited. The information helps the website know certain things such as when you last visited the site, what you purchased, your name, and your address and can be used for authentication purposes. Cookies also are used to develop profiles of users online activities over many years. In a blog post on the Electronic Frontier Foundation titled, "How Online Tracking Companies Know Most of What You Do Online (and What Social Networks Are Doing to Help Them)," the author, Peter Eckersley, describes how websites will include little bits of code stored in cookies that will allow third parties to collect information about you.[15] He cited an example of a user going to the job search site, CareerBuilder.com, where her information and job searches got collected by ten other companies, and she had no control over how they used that data. Eckersley recommends that users manage their cookies carefully such as setting your computer not to save any cookies after you close your browser. All the major browsers allow you restrict the cookies on your computer. You can do it yourself in the settings section or get help from a tech center or local computer shop if you are not comfortable changing settings on your computer. Here is where you need to decide how important your privacy is to you. Disabling and restricting cookies in your browser will help protect your privacy. However, some of the conveniences in web browsing and shopping will not work if you disable cookies on your machine or smartphone. Some such conveniences are a site remembering you and helping to

---

[15] Peter Eckersly, "How Online Tracking Companies Know Most of What You Do Online (and What Social Networks Are Doing to Help Them)," *Electronic Frontier Foundation*, September 21, 2009; accessed on line April 2, 2017

autofill some of your information, remembering what you bought before, and recommending related items. Here is the sticking point. In the case of personal action, you need to balance your comfort with your principles. Are you willing to have a less enjoyable internet experience in favor of protecting your privacy? Apart from cookies, there are many other ways you leave a digital trail for others to use to build a profile of you.

On April 3, 2017, the President of the United States signed a bill to remove the restrictions on internet service providers like AT&T, Comcast, and Verizon to collect and sell your internet search history without your consent.[16] Internet service providers (ISPs) make your access to the internet possible, and since everything you do on the internet from watching YouTube videos to shopping and even checking your medical records passes through your ISP, they hold a perfect position to gather detailed information which marketing firms will eagerly gobble up. On the darker side, hackers and thieves may use this type of information to steal your identity or even determine when you will be on vacation, leaving your home vulnerable to theft. Now, there is a way to protect yourself online called a *virtual private network*, or VPN. A VPN creates a private network like the one you would have in an office where all your computers are connected safely inside your company without access to the outside world. A VPN acts as a private network but works across a public network such as the internet.

---

[16] S.J.Res. 34: A joint resolution providing for congressional disapproval under chapter 8 of title 5, United States Code, of the rule submitted by the Federal Communications Commission relating to "Protecting the Privacy of Customers of Broadband and Other Telecommunications Services," govtrack, https://www.govtrack.us/congress/bills/115/sjres34,

The virtual private network uses encryption and private computers to mask what you are sending and receiving on the internet and your identity. Without a VPN, when you click on a website, your computer provides its address, like the address on your home, so that the information you clicked finds its way back to your computer. It is the same as having a return address on normal mail. Your request goes out, and the results get mailed back to your computer. Now the traffic to and from your computer can be recorded by other parties such as Google and Microsoft as well as your internet service provider. The VPN, instead of giving out your address when you click on a website, gives out the address of one of the VPN provider's private servers. The information comes back to the VPN server and is then encrypted and securely sent back to your computer. The internet service providers and other groups cannot detect what you are clicking on or who you are. Using a VPN significantly increases your privacy online; hackers and thieves cannot see your data or who you are. There are many VPN providers that charge for the service, or if you like configuring your computer yourself, you can set up a VPN using public servers. The VPN represents a robust method for protecting your privacy when using your computer, phone, or tablet.

The balance between personal protection and the allure of new technology presents itself clearly in the light of the new smart speakers such as Amazon Echo and Google Home. Smart speakers are small, discrete devices designed to sit in your home in an *always on* mode. They are connected to computers at Amazon or other providers and are recording what is happening in your home twenty-four hours a day. When you address the speaker by its name such as Alexa for the Amazon product, it springs to life using advanced voice recognition to understand your requests. The user can ask for Alexa to play a particular song, turn on the news, or recall a favorite recipe. For elder care, the system can be used to call for help if something has happened like a fall or other injury. For Echo at home, Siri on your phone, or Cortana on your computer to work, it needs to listen to and record your voice to learn how you speak to be ready for your next question or command. All of this information gets recorded, so the computers at Amazon and Google can learn to understand you better. Such continuous monitoring opens the door to people's homes and threatens their privacy.

Recently, the police subpoenaed Amazon Echo for the first time as a witness in the murder of Victor Collins on November 22, 2015, in Bentonville, Arkansas. Mr. Collins was found dead in the hot tub at the home of James Andrew Bates.[17]   The men, along with a third man, had spent the evening drinking beer and vodka and watching a University of Arkansas football game. Mr. Bates claims to have gone to bed early, leaving the other two in the hot tub only to awake in the morning to find Collins dead in the hot tub. The investigation surmised that Mr. Collins died by strangulation, and the owner of the home received the charge of first-degree murder. Apart from the tragedy of the case, the investigation is remarkable because it took Mr. Bates

---

[17] Terri Osborne, "'Alexa, Who Did It?': Arkansas Police Subpoena Amazon Echo Device In Ex-cop's Hot Tub Murder," Crimefeed.com,

Amazon Echo as evidence. Investigating officers noted an Echo in Mr. Bates home. Knowing that Echo is in an always-on mode, the police subpoenaed Amazon to release any audio from that night with hopes that it could shed light on the case. Amazon subsequently on February 17, 2017 filed a motion to reject the police subpoena because it violates consumer privacy rights. Amazon went on to claim that the state needs to prove that it shows a heightened need for the evidence to compel the company to release the records. Later, Amazon complied with the request to hand over the data when Bates (the defendant) volunteered to give the evidence to the prosecution. Even with the Amazon data, though, the prosecution requested the judge to drop all charges when they determined more than one reasonable explanation for the death. In late 2017, a judge dismissed the case against Bates due to a lack of clear evidence linking the death to him.[18] The importance of this case is that it shows your Echo is recording everything in your home twenty-four hours a day, and that the recording can be subpoenaed, shared by Amazon, and used as legal evidence in the court of law.

Keep Good Records

Banks today, as the place where people store money and get credit, began in Medieval Europe with the Knights Templar. The Knights Templar was an order of warrior monks founded in 1119 and was active

---

December 29,2016. Accessed online April 9, 2017
[18] Nicole Chavez, "Arkansas judge drops murder charge in Amazon Echo case," *CNN*, December 2, 2017, accessed online March 14, 2018

for nearly 200 years. The Knights Templar was closely related to the crusades and provided protection and financial infrastructure for the thousands of Christians making pilgrimages from all over Europe to Jerusalem in the Holy Land. A pilgrim leaving from London would have to travel over 3,000 miles to get to Jerusalem. They had no way to carry enough provisions to cover such a great distance and carrying enough gold and jewelry to pay for provisions along the way made the pilgrims good targets for bandits. The Knights Templar, apart from fighting and protection, also devised a system of banking where a pilgrim could leave an amount of money with the Knights Templar, for example, back in London at the beginning of the pilgrimage. In exchange, the Knights would provide a letter of credit stating how much money they can receive from the Knights Templar strongholds all the way to Jerusalem. Essentially, the Knights Templar created the first debit card. Later, Italian bankers such as the Medici Family perfected the issuance of credit loans with interest that would allow merchants to travel without money and acquire local currency using the Medici banking system that stretched across Europe. In essence, banking is the same today. It is based on trust and exchange. People trust banks to keep their money safe, and banks offer credit and assure that financial transactions are legitimate between the sender and receiver.

Today, many financial transactions happen electronically. When banks transfer money from one bank to another, the banks talk to each other through computers over the phone or on the internet. The day of the paper statement is dwindling as more banks are pressuring their clients to give up their paper bank statements for electronic records only accessible through the internet. Although convenient, there are some downsides to not keeping paper records of your banking. In an unusual report from the National Consumer Law Center titled, "Paper Statements: An Important Consumer Protection" by Chi Chi Wu and Lauren Saunders determined that paper statements serve to safeguard customers better from hidden fees and unscrupulous conduct.[19] "Consumers who see their statements are more likely to notice if they have been subject to fees or charges they did not authorize or expect, or that were far more expensive than anticipated." The authors point out that not everyone has access to the internet, and even if they do, the small screen of a smartphone is harder to examine than a paper statement. Moreover, they cite a case study that found electronic statements can get lost in large email files, resulting in missed payments.

Paper records also provide evidence of the existence of an account in the case of data corruption due to cyber-attacks, the loss of power, or a natural disaster. According to Russel Brandom at *The Verge*, in an article titled, "A New Ransomware Attack Is Infecting Airlines, Banks, and Utilities Across Europe," cyber-attacks compromised banking records in Ukraine.[20] Cyber-attacks are getting more sophisticated, and files can get corrupted or even encrypted so that the bank cannot read any of its records. We have seen natural disasters such as Hurricane Maria that devastated

---

[19] Chi Chi Wu and Lauren Saunders, "Paper Statements: An Important Consumer Protection," *National Consumer Law Center*, March 2016, accessed online November 4, 2017
[20] Russel Brandom, "A New Ransomware Attack Is Infecting Airlines, Banks, and Utilities Across Europe," *The Verge*, June 27, 2017, accessed online November 4, 2017

Puerto Rico, leaving many without power for weeks, making electronic banking nearly impossible. Massive solar flares can also corrupt electronic records on a massive scale. Back in 1859, a solar superstorm hit the world and took out the telegraph system and, in some cases, caused the telegraph machines to spark and even burst into flames. Experts predict that a solar superstorm today would take out many computers and perhaps damage electronic records.[21]

Using paper money and credit statements began nearly a thousand years ago with the Knights Templar as a way to help travelers not have to carry gold and jewels that would make them targets for thieves. Today we still use banks to safeguard our money and obtain credit. As banks look to reduce the cost of mailing paper statements, they are pushing consumers to forgo the paper statement for all electronic statements. Consumers should not give up their paper statements so readily. The paper statement serves a number of uses such as a helpful reminder of payments due to providing physical evidence of your account or as a safeguard in the case of a cyber-attack or natural disaster that corrupts the computers and networks serving them.

As a user of any smart technology that listens to and records your private life, from Amazon Echo to Cortana or Siri on your laptop or phone, you need to be aware of the depreciation of the privacy in your home, your office, or your car. You have options to erase some of the information collected by these devices, but the truth is that apart from muting the microphone on your

---

[21] Richard A. Lovette, "What If the Biggest Solar Storm on Record Happened Today?" *National Geographic News*, March 4, 2011, accessed online July 14, 2017

smart speaker, you have opened your home to the outside world in an unprecedented way. Efforts to protect our jobs, our safety, and our privacy continue to emerge at every level of society from self-regulating computer science researchers and government regulations to legal precedent. Additionally, people can perform personal actions such as setting up a VPN or muting their smart speaker to help manage the rapidly evolving world of artificial intelligence, machine learning, and robotics. With every advance in technology, we realize unique and exciting benefits, but everyone needs to consider the costs and not just take what comes next.

Chapter 10
The Future

I would gladly risk feeling bad at times, if it also meant that I could taste my dessert.

—Commander DATA, *Star Trek: The Next Generation*, "Hero Worship"

Narrow AI

The future of AI will bring great success and riches to some and prove deleterious to others, and AI will change the way markets work and how we view employment. Throughout this book, we have looked at different types of AI and robotics in light of how to profit and protect yourself. We looked at examples of machine learning and deep learning and how these data-hungry algorithms are transforming specific areas of business, research, society, and even our personal lives. Much of the examples and discussion has focused on the near term to understand this emergent technology. Moreover, most of the AI that we have looked at is categorized as narrow AI, or weak AI. Narrow AI describes the AI that does one type of task very well such as facial recognition, translation, speech recognition, prediction, and planning. However, narrow AI cannot change its mind and work on some other problem. A person effortlessly shifts from planning a party to talking with a friend to attending to a dispute among their children. People apply their intelligence freely to issues at hand as they arise. People possess general intelligence. General intelligence in the world of computing is called *artificial general intelligence*, or AGI. AGI remains an aspiration for AI researchers, and many believe it to be the real future of AI. Before we finish the book with a look down the road, we must address the growing power of artificial intelligence to affect our lives and even manipulate how we view the world and make decisions for ourselves, our families, and our society, including how we select our employment to how we perceive the truth.

Various consultants, futurists, and pundits provide a view of the future that looks very different from today. Lori G. Kletzer, professor of economics at Colby College and University of California Santa Cruz, notes in an article in the *Harvard Business Review* that, "Currently, less than 5% of occupations are entirely automated, and about 60% of occupations have at least 30% of tasks that can be automated."[1] An article in Forbes states a more foreboding number, "Statistics say that 47% of all employment opportunities will be occupied by

---

[1] Lori G. Kletzer, "The Question with AI Isn't Whether We'll Lose Our Jobs — It's How Much We'll Get Paid," *Harvard Business Review*, January 31, 2018, accessed online April 4, 2018

machines within the next two decades."[2] Some jobs will completely disappear, but in other cases, only specific activities that people do at work today will face automation. Such an expansive view of the penetration of artificial intelligence and robotics into every aspect of our lives indicates more disruption to come. People do adapt, and the shifting roles of people in the market will emerge with specific human skills still in demand such as physical skills in complex environments, mental skills in problem-solving, and social skills in some interpersonal settings from negotiations to sales and health care.

Creativity

Creativity, as discussed throughout the book, remains as one of the essential human qualities that machines may exhibit at a low or narrow level. AI may be able to imitate the styles of artists, musicians, and writers, but that is not the same as becoming the next great artist, musician, or writer. Storytelling remains a foundational form of human creativity that certainly predates the invention of the written word. In computer science, storytelling falls into a category known as creative linguistics. Researchers in the area of computer-generated storytelling have tried to develop systems since the 1970s that attempted to treat storytelling like a problem with a series of steps that need completing until a solution arrives. Many of the systems try to compose stories using formulas. At the most basic level, sports reporting already uses

[2] Harold Stark, "As Robots Rise, How Artificial Intelligence Will Impact Jobs," Forbes, April 28, 2017, accessed online November 12, 2017

computers to write recaps of baseball, basketball, and other events.

The robo-journalist does not merely work right out of the box. A content editor needs to train the computer in style and content. Headlines from Japan touted a novel written by artificial intelligence that passed the first round of a literary contest known as the Nikkei Hoshi Shinichi Literary Award with over 1400 submissions.[3] Upon further investigation, the story was not written actually from scratch by the computer. Instead, a team of writers helped to frame the story and, with guidance, let the AI rearrange the narrative. The point is not to put down the AI-generated stories, but rather to point out some strategies of how to benefit from such emerging technologies.

Man joining forces with artificial intelligence in the creative storytelling process and more generally in games and problem solving has been coined *centaurs*. The term comes from a reference to the mythological creatures with the head, upper body, and arms of a human and the lower body and legs of a horse. Centaurs in computer science refer to teams that use the strengths of AI and computing to augment the powers of the human intellect. In a new type of chess called *centaur chess*, or cyborg chess, man and machine team up to compete in tournaments that produce better chess results than man or machine alone. Garry Kasparov, the famous world champion chess master, infamously lost his title to IBM's Deep Blue artificial intelligence computer in 1997. Deep Blue's victory marked the

---

[3] "This Japanese Novel Authored by a Computer Is Scarily Well-Written," *Fast Company*, March 28, 2016, accessed online August 10, 2016

first time a world champion chess player lost to a computer. Over the next twenty years, Kasparov has become interested in artificial intelligence. Despite his well-publicized loss, he has envisioned a new future where man and machine work together augmenting human and machine skills. Kasparov discussed man's future with artificial intelligence in a book titled, *Deep Thinking: Where Machine Intelligence Ends and Human Creativity Begins*.[4] He suggests a future with machines augmenting human capabilities and in as yet unknown industries and unexplored frontiers.

Bias

Bias hurts when it negatively affects your life. Negative bias in hiring, receiving a loan, getting accepted to a school, being diagnosed at the doctor's office, and such robs people of their dignity and a feeling of equity in the world. Research into human bias over the past forty years has revealed a variety of biases that people exhibit without even thinking about it. One type of bias called *availability bias* described in research of Amos Tversky and the Nobel Prize winning economist Daniel Kahneman, first published in 1973, specifies the mental short cuts people use when faced with a decision[5]. Their research showed that people would make a judgement about someone based on how easily they remember people in a similar situation. One example the authors use in their publication relates

---

[4] Garry Kasparov, *Deep Thinking: Where Machine Intelligence Ends and Human Creativity Begins*, PublicAffairs, 2017
[5] Amos Tversky and Daniel Kahneman, "Availability: A Heuristic for Judging Frequency and Probability," *Cognitive Psychology*, 5, 207-232 (1973)

to a mental process of predicting if a couple will get a divorce. The research showed that if the couple in question looks or acts similar to another couple that you remember that had a divorce, you will judge a divorce to be more likely for the married couple, but if you do not easily recall such a memory, you will not predict divorce for the couple. In other words, we tend to judge based on how easily we remember a similar situation, which may not be genuinely relevant at all. Another type of bias called *confirmation bias* refers to people's tendency to select evidence that supports the beliefs they already hold. For this reason, the discussion about *fake news* remains such a vital topic. The tendency for people to believe and share information that supports their beliefs even in the face of conflicting information sits at the heart of the influence of social media driving public opinion. Hunt Allcott and Matthew Gentzkow, professors of economics at New York University and Stanford University respectively, published an article titled, "Social Media and Fake News in the 2016 Election."[6] In their report, they note how groups will share stories that support their beliefs, creating an information bubble, or *echo chamber*. Social media supports the echo chamber by offering similar or supporting information among friends and like-minded people.

Computers often have a reputation for being cold calculators, devoid of bias. In many respects, the reputation fits. When a computer performs calculations,

---

[6] Hunt Allcott and Matthew Gentzkow, "Social Media and Fake News in the 2016 Election," *Journal of Economic Perspectives,* Volume 31, Number 2, Spring 2017, Pages 211–236, accessed online April 7, 2018

without question it produces the same results every time. It will not vary. However, computers used for artificial intelligence can produce biased results. Anuranjita Tewary, the Chief Data Officer at Mint and the founder of the information technology company Level Up Analytics, found when she worked at the social networking company LinkedIn, their algorithm displayed high salary jobs much more frequently to men than to women[7]. She determined that the developers of the algorithm and most of the initial users were men, which trained the program to favor men or be biased toward men. One of the key elements to artificial intelligence is that artificial intelligence centers on the fact that the computer needs to learn, and it learns much like people do, by repetition. For example, to train a facial recognition program, the computer needs to analyze tens of thousands, even millions of faces to learn how to recognize a face. Here is where bias can creep into an artificial intelligence program. If the faces used for the training do not represent the population, the algorithm will display a bias to what faces it can recognize. A graduate researcher at the Massachusetts Institute of Technology Media Lab, Joy Buolamwini, along with Timnit Gebru of Microsoft, analyzed several facial recognition programs from companies such as IBM, Microsoft, and Face ++ to determine if the different systems showed bias. She found that all the programs showed a bias in race and gender determination[8]. Moreover, the algorithms failed most often when attempting to classify darker-

---

[7] Hope Reese, "Bias in machine learning, and how to stop it," *TechRepublic*, November 18, 2016, accessed online April 6, 2018
[8] Joy Buolamwini and Timnit Gebru, "Gender Shades: Intersectional

skinned females (up to a 34% failure rate). The facial recognition algorithms performed best on categorizing light-skinned males which also correlates with the high proportion of light skinned-male face images in the training set. Such results should make any group from businesses to government and academic institutions take a very hard look at data selection when training an artificial intelligence system to combat bias in the results.

Research into human thinking uncovered a number of biases in people, including selection bias and confirmation bias. Such biases can lead to inappropriate conclusions and inequitable decisions. Although computers often hold the reputation of being cold and unbiased, research demonstrates the artificial intelligence systems too can become biased unless great care goes into selecting the data that trains the system and analysis of the results to determine if bias has worked its way into the system. Microsoft learned the hard way that uneven human input can make an algorithm biased in a very bad direction when they released their twitter-based chatbot named Tay[9]. In less than twenty-four hours after launch, Tay needed to be taken down because it had learned from specific tweets to become a racist and a misogynist. Properly designing the AI and feeding it the right information remains critical to reducing bias and making AI that works for everyone.

Accuracy Disparities in Commercial Gender Classification," Proceedings of Machine Learning Research 81:1–15, 2018, accessed online April 7, 2018

[9] Sarah Perez, "Microsoft Silences Its New A.I. Bot Tay, After Twitter Users Teach It Racism [Updated], *TechCrunch*, March 24, 2016, accessed online September 12, 2017

Robots

A robot may not injure a human being or, through inaction, allow a human being to come to harm.
A robot must obey the orders given it by human beings except where such orders would conflict with the First Law.
A robot must protect its own existence as long as such protection does not conflict with the First or Second Laws.

The Three Laws of Robotics
"Runaround," 1942
By Isaac Asimov

The field of robotics, driven by an astonishing number of engineering breakthroughs, has produced remarkable humanoid robots that can walk on two feet, perceive their world through sight, sound, and touch and even interact with people at a social level. Honda's ASIMO, the world's most famous humanoid robot, first introduced to the public in 2000, walks, greets people, responds to questions in multiple languages, and even conducted the Detroit Symphony Orchestra in 2008. More advanced robots continue to emerge in part due to advances in engineering to make the robots lighter and stronger. In addition to the engineering marvels that give us robots of all shapes and sizes, artificial intelligence has monumentally contributed to the development of more lifelike and complex robots than ever before. In fact, some robots are interactive enough that researchers already must consider how to teach robots right from wrong.

Isaac Asimov famously proscribed his Three

Rules for Robotics in a short story called "Runaround." In many ways, these rules should apply to robot development today. Work at Tufts University's Human-Robot Interaction Lab by researchers Gordon Briggs and Matthias Scheutz demonstrates how robots can be programmed to evaluate commands from humans. They describe their work in a paper titled, "Sorry, I Can't Do That: Developing Mechanisms to Appropriately Reject Directives in Human-Robot Interactions."[10] The paper describes how they developed a reasoning algorithm that helps a robot evaluate verbal commands from people. Their reasoning system empowers the robots to deny commands that would harm the robot or would harm something or someone else. Briggs' and Scheutz's algorithm gives a robot the tools to evaluate the potential results of any action it is commanded to perform in light of the environment that the robot encounters. An example in the paper describes an actual scenario where a human asks a small robot named Dempster to walk to the edge of a table, and the robot replies, "Sorry, I cannot do that as there is no support ahead." The individual commands, "Walk." Dempster then replies, "But, it is unsafe."[11] The sequence of commands and responses shows impressive interactivity between the robot and the researcher, and moreover, a natural dialog between a human and a robot empowered to protect itself.

Not all questions of right and wrong are clear-cut, and robots will be asked to make more nuanced

---

[10] Gordon Briggs and Matthias Scheutz, "'Sorry, I can't do that': Developing Mechanisms to Appropriately Reject Directives in Human-Robot Interactions," Association for the Advancement of Artificial Intelligence, 2015
[11] Ibid

decisions as they become more sophisticated and independent. Given the ability of artificial intelligence to learn through experience, Mark Riedl and Brent Harrison, researchers at the Georgia Institute of Technology, developed a system called Quixote that takes a different approach to teaching robots how to make the right moral choice when confronted with a situation. Using the observation that human children learn right from wrong from stories and from being punished for making the wrong choice, Quixote reads children's literature and also has a system of reward and punishment to learn from the decisions it makes. Such a learning style is known as *reinforcement learning*, where the machine is designed to remember actions that produce a positive effect and use these learnings to make informed decisions in the future.

The rise of robots and artificial intelligence continues to produce more sophisticated humanoid robots that can make their own decisions and act independently of human operators. Robots continue to acquire more autonomy, prompting researchers to develop ways for robots to make right choices that respect Asimov's Three Laws of Robotics and are consistent with human morals and ethics. A group of research institutions, including Tufts University, Rensselaer Polytechnic Institute, and Yale University, received a Multidisciplinary University Research Initiative (MURI) from the United States Department of Defense titled, "Moral Competence in Computational Architectures for Robots." The initial attempt through research is to understand human moral and ethical actions, adopt human morals and ethics to human-robot interactions, and to develop computer programs that will enable robots to act morally and ethically.[12] Thomas Arnold and Matthias Scheutz from the Human-Robot Interaction Laboratory at Tufts University have written on the challenges developing the proper behavioral models for soft robotics.[13] They point out the difficulties of making robots that respond to and communicate through the very basic sense of touch. The authors looked at the ethical consideration of developing sex robots. Summarizing their research in the social perception of sex robots, the analysis determined that careful design and more significant social implications need to be considered as the field of robotics evolves. Developers need to consider how people bond through touch and help avoid misleading and counterproductive human-robot attachments forming.

Society needs to carefully consider how it will be impacted by the frightening prospect that people will bond at the most fundamental level with robots. One group called Campaign Against Sex Robots strongly asserts that sex robots pose an imminent threat to society by further objectifying women and children.[14] The organization also proposes "that the development of sex robots will further reduce human empathy that can only be developed by an experience of mutual relationship." The group cites work by Kathleen Richardson of De Montfort University in the UK that draws connections between the objectification in

---

[12] "Moral Competence in Computational Architectures for Robots," https://hrilab.tufts.edu/muri13/, accessed online November 11, 2017
[13] Thomas Arnold and Matthias Scheutz, "The Tactile Ethics of Soft Robotics: Designing Wisely for Human–Robot Interaction," *Soft Robotics*, 4, 2, 81—87, 2017
[14] Campaign Against Sex Robots, https://campaignagainstsexrobots.org/about/, accessed online December 16, 2017

prostitution and the further objectification of women in sex robots. Religious groups prohibit sex with robots because in their view sex is for procreation and bonding between spouses. Other segments of society take a different view and see a future of humans having sex with robots. In a Foundation for Responsible Robotics Consultation Report titled, "Our Sexual Future with Robots," the authors touch on some important issues around sex robots.[15] They summarized a number of surveys about whether a person would potentially have sex with a robot. The surveys taken over the past several years found a range of responses with men more likely, up to 60%, to consider having sex with a robot and women significantly less likely, around 9%. A 2017 survey conducted in the United States by YouGov, an international market research and analytics firm, found that 25% of American men would consider having sex with a robot while only 13% of women would consider it. The YouGov survey also included the question of whether US adults would consider if their partner had sex with a robot to be cheating. According to the survey, 32% said it would count as cheating, 33% said it would not, and the rest were not sure.[16] The future is here as companies such as RealDoll commercially offer robotic sex dolls that respond to touch and use artificial intelligence to hold conversations. Proponents of companion sex robots such as the founder of RealDoll and its parent company Abyss Creations, Matt McMullen, sees a very positive role in society for sex

---

[15] Noel Sharkey, Aimee van Wynsberghe, Scott Robbins, and Elanor Hancock, "Our Sexual Future with Robots," July 5, 2017, https://www.scribd.com/document/353020582/Report-Our-Sexual-Future-With-Robots#from_embed, accessed online December 16, 2017
[16] "1 in 4 men would consider having sex with a robot," today.yougov.

robots. In an interview with Pam Kragen from the *San Diego Union Tribune*, McMullen, "sees Harmony [a sex robot sold by RealDoll] more as a comforting conversation companion like Apple's Siri, albeit one capable of having sensual conversations and telling naughty jokes." The topic should not be taken lightly, and society should not be merely asking can these robots be made, but *should* they be made? With robots edging closer to being genuinely conscious machines, we need to find ways for the machines to help and support society and not become a liability to the world.

In "Moral Robots," Matthias Scheutz and Bertram F. Malle note that humans have developed morality and ethics, and through observation and training, we learn, share, and apply our moral code in society. Moreover, we must encode morality into increasingly autonomous robots. "Hence, these new social robots will require higher degrees of autonomy and decision-making than any previously developed machine, given that they will face a much more complex, open world. At these levels of autonomy and flexibility in near-future robots, there will be countless ways in which robots might make mistakes, violate a user's expectations and moral norms, or threaten the user's physical or psychological safety. These social robots must therefore also be moral robots."[17] Part of the challenge of developing a moral robot will be giving the robot the ability to act morally in a situation where it must also consider protecting itself. Chapter 3 mentions that Sea Hunter, a diesel-powered ship, was designed to navigate the waters of the world

com/news/2017/10/02/1-4-men-would-consider-having-sex-robot/, September 26, 2017, accessed online December 16, 2017
[17] Matthias Scheutz and Bertram F. Malle, "Moral Robots," In

on its own without the close support of other ships or people. *Sea Hunter* is a 132-foot-long diesel-powered trimaran that holds enough fuel for it to spend months at sea hunting submarines without human intervention. Military drones will need to make quick decisions at times out of any contact with human commanders to complete a mission or to protect itself from the hostile action either by enemy humans or robots. The trend apparently leads to more, not less autonomy for robots in the future.

Artificial General Intelligence

For most of the book, we have looked at examples of narrow artificial intelligence, which refers to computer algorithms that do a particular job or process that previously only people could do. Narrow artificial intelligence recognizes faces, learns to translate languages, guides vehicles safely through streets, and more, but it does not have general intelligence like a person. Artificial general intelligence goes beyond the machine learning and deep learning that we have been looking at so far. AGI, once it is developed, would be an entirely independent, *conscious* entity, fully capable of surviving and thriving on its own, an entity with its own will and motivations. Researchers, ethicists, politicians, and concerned citizens have worked to define rules and guidelines for the development of beneficial artificial intelligence. In a letter written by the Future of Life Institute titled, "Research Priorities for Robust and Beneficial Artificial Intelligence: an

K. Rommelfanger and S. Johnson (eds.), Routledge *Handbook of Neuroethics*, Routledge/Taylor & Francis. 2017, New York

Open Letter."[18] The letter acknowledges the potential benefits of super-intelligent machines to help solve some of man's most significant problems such as disease, poverty, and hunger as well as noting the real danger of AI in some sectors.

The critical question is: can we regulate our progress at all?

In the "There Was an Old Lady Who Swallowed a Fly" song, the old lady goes on swallowing ever larger and larger animals, including a cat, a dog, a goat, and a cow to catch the previous creature she had swallowed. The song reminds us that the most recent remedy may, in fact, kill her. At the end of the song, she swallows a horse, and rather abruptly, she dies. The song in a humorous way tackles the issue of humans creating a new problem by trying to fix an existing problem with some scientific, mechanical or biological solution. A classic example, as in the song, is a biological control— the introduction of a predator species to control a pest or a disease-carrying animal or plant. Familiar types of biological control can be found in agriculture and form a vital part of integrated pest management. Take for example the domestic use of ladybugs for controlling aphids in the garden or using goats to remove poison ivy.

Not new, biological control dates back thousands of years. Chinese literature from around 300 AD mentioned the sale and use of yellow ants for the protection of citrus crops. Since then, numerous

---

[18] Stuart Russell, Daniel Dewey, and Max Tegmark, "Research Priorities for Robust and Beneficial Artificial Intelligence, Association for the Advancement of Artificial Intelligence," futureoflife.org, Winter 2015, accessed on line March 19, 2017

examples of biocontrol fill agricultural research from the introduction of predators like the ladybug to parasitoids that lay their eggs in or on another species. For example, parasitic wasps like the chalcid wasp lay their eggs in an insect pest known as the whitefly. After the eggs hatch inside the host, the developing baby wasps grow inside the host, weakening and eventually killing it. I know this sounds like something out of the movie *Alien*, but gardeners and greenhouse managers employ parasitic wasps regularly for pest control. Other types of biocontrol used today by farmers include bacteria and fungi.

Biological control offers farmers and groundskeepers alternatives to chemical pesticides that when applied correctly prove very useful. On the other hand, biological pest control has not always had its desired effect. In Hawaii, in the late 1800s, sugarcane farmers struggled with rats eating their crops. Farmers in Hawaii had heard the sugarcane farmers in Jamaica had used mongoose to control the rats in their sugarcane fields, and they released over 100 mongooses in their fields in 1883. Regrettably, no one really thought this through. There was one massive miscalculation. Rats are nocturnal; they feed and move about at night. In contrast, the mongoose hunts and moves about during the day. Needless to say, the mongoose posed no threat to the rats who continued to safely eat the sweet sugarcane at night while the mongoose spread across the island, eating up native bird species and other indigenous animals. The moral of the story is people genuinely have to think hard and long before they take irreversible action.

The rise of Zika virus in Brazil, its detection in parts of Florida, and its connection to increased birth defects has many people very concerned. Two types of mosquito transmit Zika—*Ae. aegypti* and *Ae. albopictus.* The prospect of Zika is frightening, and no vaccination has been developed yet. However, scientists recently used advanced genetic engineering to produce genetically modified male mosquitos that, when they mate with wild females, produce no young. Trials indicate this method will significantly lower the mosquito population. But what could be the other implications of no mosquitos? We must think of the mosquito in the larger ecological picture. Mosquitos don't just bite us; they serve as a food source for other animals such as fish and as population controls by spreading diseases to other animals. What might happen if they were all of a sudden extinct due to genetic modification?

Once free in Hawaii, the mongoose could not be called back and has had a lasting effect on their ecosystem. Fortunately, we now have the Food and Drug Administration and the Environmental Protection Agency to rigorously test the impact of new pest control measures on human health and the environment before they can be released. At the same time, it is my strongest belief that everyone should get to know our emerging technologies and ponder their implications. Speak up if you think something is wrong or that maybe the agencies did not do all their homework. Social media such as Twitter and Facebook offer great places to bring attention to a situation in which a company or government agency needs to be called out for their actions. Furthermore, government rulemaking must allow for public comment, which includes electronic means to communicate.[19]

How Would AGI Work and Survive Free in the World Among Us?

"Economics is the study of the use of scarce resources which have alternative uses."

Lionel Robbins
British economist

Lionel Robbins' definition marvelously captures the essence of economics in a single sentence. So often journalists, researchers, and marketers reflect on the effect of robotics, artificial intelligence, and automation on current jobs and the displacement of humans from the labor force. However, there is another aspect of the rise of AI and robots that deserves more investigation—how will people react to robots integrating into society? Academics, industrial researchers, and engineers continue to build robots that look and move like people. Additionally, advances in artificial intelligence continue to give robots more human-like qualities and more sophisticated reasoning. Ultimately, robots will have more and more autonomy, which gets back to the question of economics.

In the future, people, society, and economic systems will come to face the challenge of autonomous robots in our communities and as players in our economic system. If economics deals with the use of scarce resources, independent, self-sufficient robots will have

---

[19] https://www.regulations.gov/

to live under the same financial pressure as people do. A genuinely autonomous robot participating in society would need to support itself with energy, shelter, repairs, and upgrades. If it were truly independent, it would need to compete for scarce resources with people and other robots. Take for example the valuable minerals required to build different types of electronics such as fiber optics, visual displays, lasers, diodes, computer memory, batteries, and more. Some metals such as gold and platinum are rare and expensive and used for many purposes, including manufacturing and jewelry. Other metals called rare earth minerals with exotic names like europium and samarium are used in computer and robot manufacturing and maintenance and are also expensive due to their scarcity and already high demand. For example, lanthanum is a material for rechargeable batteries and is essential for robot function. Competition for rare earth metals, as the population of robots continues to grow, would put pressure on markets and prices and perhaps create a situation where robots could not afford the necessary parts to keep themselves going. Currently, China controls about 95% of the world's rare earth elements production and has made moves to set export quotas and fix prices.[20] Additionally, China continues to develop industries domestically that will benefit from the supply of these elements, further threatening the available export of these elements. The control of rare earth elements production by China serves as an example of resource control and begs the question of

---

[20] Jeff Nesbit, "China's Continuing Monopoly Over Rare Earth Minerals," *US News and World Report*, April 2, 2013, accessed online December 3, 2017

what may happen next. Any resource from energy, raw materials, or information may be an issue when supplies become scarce.

Throughout history, scarcity has driven conflict. The stability of world markets, trade routes, and peace have always depended on the threat of overwhelming violence to protect the peace. For the sake of completeness, the logical outcome of rising artificial intelligence embodied in robots suggests a conflict at some point. That battle will most likely be violent and intense. History tells the story over and over again about the rise of powers from warlord strongholds to kingdoms to advanced city-states and nations and eventual collapse of powers and decent back into anarchy and the rise of the warlord. Such a grim cycle needs to be contemplated when envisioning our rush to develop artificial intelligence. If the predictions of Kurzweil and other futurists of the coming of artificial general intelligence that will far outpace human intelligence come true, humans at some point will be the problem competing for scarce resources. The competition may at first play out in the marketplace with robots slowly integrating into society in places like the workplace, the military, and remote or dangerous locations.

More than just competition for minerals and energy, robots will need to compete for information and jobs in a functioning marketplace. Currently, artificial intelligence learns from massive data sets compiled from different public and private data sources such as data harvested from the web in blogs and publications as well as data gathered from customers at tech giants like Facebook and Google. The information may not be open and available, creating imbalances in artificial intelligence for robots. Already, corporations guard their proprietary information; governments protect many types of information from individual and corporate tax returns to personal health information and private citizens safeguard their personal information from identity theft. Information flow and access are and increasingly will be a crucial asset and source of conflict in the future with artificial intelligence. As for jobs, can one imagine robot resumes and work experience? Certain skills in data analysis and decision making will be strong suits for robots, so perhaps more creative job skills will still demand more human employees. In fact, the competition for employment has already begun; we already see these kinds of changes in occupations such as factory work and clerical work.

Moreover, the rise of robots creates competition for space. Physical living space and location will pit people against robots in the congested city centers. At first, people may react in different ways to robots entering their communities, creating a possibility for robot-only communities in towns and city centers. There may be legal implications and battles over whether or not landlords can refuse robots tenancy merely because they are robots. Moreover, shopkeepers may not allow robot customers, or the government may not extend privileges to robots previously given to humans. These questions will become more involved as resources begin to grow scarcer.

From an economic perspective, there will be competition whenever there are alternative uses for scarce resources. The evolution of autonomous robots has already begun with the most significant advances already happening in autonomous vehicles and cognitive computing such as IBM's Watson. Science fiction suggests various future worlds that deal with intelligent robots as enemies, but another possible picture of the future shows a slow integration of humanoid robots into society. People may grow to like and relate to intelligent robots as well as compete with them for scarce resources. The new economics will create exciting new markets that will include your robot neighbors.

However, a more violent alternative reality might be the real case. Looking at human social history as well as the broader view of evolution with the rise and fall of dominant species across millennia, the ascendance of artificial intelligence and robotics will see an advancement of power and capability that will require the displacement of humans from their current dominant role in the world. Consider that human resourcefulness, adaptability, ingenuity, and need to explore has propelled people to visit and often stay in every conceivable part of planet earth. Other species have flourished and even dominated ecosystems through the evolutionary history of the planet. Various events have caused significant changes in the animal and plant world. Researchers agree that a massive asteroid 66 million years ago colliding with earth caused an enormous cooling of the planet that killed off the dinosaurs in a massive extinction event. More recently, scientists at the University of Oregon determined that wooly mammoths, saber-tooth tigers, and other great mammals of North America became extinct about 13,000 years ago when a comet exploded over North America somewhere near where Chicago is today.[21] The explosion caused massive fires that

---

[21] Robert Mitchum, "Scientists say comet killed off mammoths, saber-toothed tigers," *Chicago Tribune*, January 2, 2009, accessed online

plunged the world into an ice age, driving many animals into extinction. Beyond the effects of natural disasters, invasive species, and disease, human migration has had a seismic impact on people, plants, and animals. The rise of a superintelligent AI may well have the same effect.

The futurist author Nick Bostrom wrote *Superintelligence: Paths, Dangers, Strategies.*[22] In his work, Bostrom warns of a superintelligence developing out of AI that would emerge unchallenged on the earth. Such intelligence will pose a threat to all humanity and would have the intelligence to defeat any retaliation from humans. His work has dramatically influenced Bill Gates, the founder of Microsoft, Elon Musk, the founder of Tesla and SpaceX, and other influential business leaders and scientists such as Stephen Hawking to publicly voice their concerns about the existential threat AGI poses for humanity.

The notion that as long as man controls the power and the means of production, we will always be able to control the robots and AI we produce, should be questioned. The dependence of robots and AI on human infrastructures such as electrical grids, supply lines, and factories to manufacture and assemble the electronic and mechanical components of robots serves as a point of control for man. Turning off the power has always been perceived as the last chance fail-safe. However, the evolution of 3D printing and the development of robots that power themselves with organic matter will change the future landscape of human control over

August 7, 2017
[22] Nick Bostrom, *Superintelligence: Paths, Dangers, Strategies*, Oxford University Press, 2014

AGI Robots.

The website Edge.org contains a remarkable dialog among some of the world's brightest and most accomplished thinkers. To drive the dialog, Edge.org poses a big question each year and collects the responses on their website. The annual question draws considered reactions from around the world and from many disciplines. The 2015 question, "What Do You Think About Machines That Think?" drew hundreds of responses and spanned the range from alarmed and cautious visions of a future where humans fall into second place behind artificial superintelligence to views that dismiss the possibility of AGI at all. The astrophysicist and Nobel Prize winner John C. Mather cautioned in his Edge.org response in unequivocal terms that machines are evolving and are subject to the same Darwinian pressures as any organism. He says, "So far we have found no law of nature forbidding true general artificial intelligence, so I think it will happen, and fairly soon..."[23] Michael Vassar, the co-founder of MetaMed Research, pondered that a super intelligent machine could lead to our extinction. He cautions too that leaving the decision making to machines for the public good may lead to authoritarianism that runs against what people think of as happiness. The Harvard Professor, George Church, starts his response with the declaration that he himself "is a machine that thinks."[24] According to some, the world of intelligent machines and even human machine hybrids will be upon us but just how soon remains the only debate.

---

[23] John C. Mather, "2015: What Do You Think About Machines That Think?," Edge.org, 2015, accessed online November 12, 2017
[24] Ibid

Other respondents to the Edge.org question: "What Do You Think About Machines That Think?" were skeptical of the likelihood of the rise of intelligent machines that could give the human race a run for its money or even make us extinct. Satyajit Das, the author of *Age of Stagnation*, showed little confidence that people can build thinking machines. He writes, "The human species is simply too small, insignificant and inadequate to fully succeed in anything that we think we can do."[25] The physicist, Freeman Dyson, flatly states that a thinking machine in the near future is not likely. The Professor of Psychology at Princeton University, Eldar Shafir, questions how a machine, if it cannot experience the range of emotions from love to fear, can be truly a thinking machine. He uses references to the decisions people must make when there are no good choices. True thinking, he asserts, occurs when having to choose between two bad outcomes. The Austin B. Fletcher Professor of Philosophy at Tufts University, Daniel C. Dennett chides *The Singularity*, the point at which AI surpasses human intelligence, as a surprisingly persistent urban legend. He further worries that far from man creating superintelligence that "The real danger is basically clueless machines being *ceded* authority far beyond their competence."[26]

On the one hand, some voices doubt that we will see the emergence of superintelligence; instead, they propose it is more likely man will cede authority to non-thinking machines that risk the weakening of human intelligence. Personally, I find the references to Darwinian evolution driving the rise of artificial

---

[25] Ibid
[26] Ibid

superintelligence to be a crucial concept to ponder. I see three necessary ingredients for developing a superintelligence: a drive to survive, need to procreate, and genuine curiosity. If such drives and curiosity can be engineered into AI, then I think that it will be just a matter of time before we see actual artificial superintelligence. Time will only tell where thinking machines will go, but everyone, not just the great minds, should engage in the debate to participate in our evolving relationship with machines.

Index

Index